Elegant Glass with Corn Flower

Imperial Candlewick, Heisey, Tiffin & More

Walter T. Lemiski

Schiffer Publishing Ltd

4880 Lower Valley Road, Atglen, PA 19310 USA

Dedication

To my beloved better half,
Kimberley, for her ever-
steadfast encouragement,
and shared joy of collecting
and love of research.

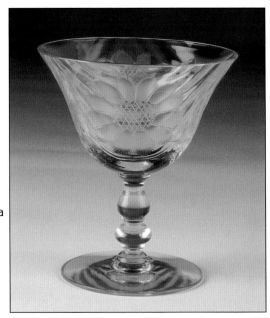

Library of Congress Cataloging-in-Publication Data

Lemiski, Walter.
 Elegant glass with corn flower : Imperial
Candlewick, Heisey, Tiffin & more / by Walter
Lemiski.
  p. cm.
 ISBN 0-7643-2141-2 (pbk.)
 1. Cut glass—Collectors and collecting—United States—
Catalogs. 2. W.J. Hughes and Sons Corn Flower
Limited—Catalogs. 3. Glass manufacture—United
States—History—20th century. I. Title.

NK5203.L46 2005
748.2'0973'075—dc22

2004016375

Designed by Mark David Bowyer
Type set in ShelleyVolante BT / Korinna BT

ISBN: 0-7643-2141-2
Printed in China
1 2 3 4

Published by Schiffer Publishing Ltd.
4880 Lower Valley Road
Atglen, PA 19310
Phone: (610) 593-1777; Fax: (610) 593-2002
E-mail: Info@schifferbooks.com

For the largest selection of fine reference books on this and
related subjects, please visit our web site at
www.schifferbooks.com
We are always looking for people to write books on new and
related subjects. If you have an idea for a book please
contact us at the above address.

This book may be purchased from the publisher.
Include $3.95 for shipping.
Please try your bookstore first.
You may write for a free catalog.

In Europe, Schiffer books are distributed by
Bushwood Books
6 Marksbury Ave.
Kew Gardens
Surrey TW9 4JF England
Phone: 44 (0) 20 8392-8585; Fax: 44 (0) 20 8392-9876
E-mail: info@bushwoodbooks.co.uk
Free postage in the U.K., Europe; air mail at cost.

Contents

Foreword

Glass, in itself, is an inanimate object, the result of a manufacturing process, an ambiguous and, in pure form, colorless lump. But give it design, form, color or decoration and it becomes alive, animated by the unknown minds and hands that have created, formed or decorated it. The chemists and designers names are lost to history; the only trace left is the objects that they created. The only detail of the person who formed, blew, or cut designs in these glass objects is the skill of their craftsmanship. Yet, the glass, which we today call "Elegant Glass," is recognizable, distinguishable, and collectible, based on its form, color, and style. It tells stories of the era and in small ways the stories of the people who created them.

In recent years, more and more collectors have wanted to learn more about the pattern names of elegant glass. They research the names of the companies that produced these products and the different colors or items originally available. Additional information adds new stories and information to their treasured possessions. This knowledge brings an increased level of familiarity and emotion to these objects.

In the case of the pieces of Elegant Glass that have been cut with the familiar W.J. Hughes Corn Flower pattern, an additional level of knowledge is possible for the collector or admirer. The beauty of the design and the skill of the cut give us a comforting sense of connection to its creator and, often, cutter, William John Hughes. W.J. Hughes was born in Amaranth Township, Dufferin County, Ontario, in 1881, the son of Henry Hughes and Margaret Jane Armstrong. He was raised in Amaranth and, later, Melancthon Township. It is believed that, after he moved to Toronto to seek employment, he designed and began cutting the Corn Flower pattern at 212 Wychwood Ave., Toronto.

From 1916 to 1951, the elegant glass that Hughes selected to decorate produced some of the rarest and most beautiful glass cut by the company. He ran the company known as W.J. Hughes Corn Flower Ltd. until his death in 1951. The company continued to cut this pattern on glass until 1988 under the direction of his son-in-law and daughter, Pete and Lois Kayser. This couple increased the market across Canada and it became known from coast to coast.

This wonderful patterned and, often, colored glass of the Depression Glass era provides a glimpse at the history of commerce and industry during that period. The glass blanks were cut in the basement of W.J. Hughes's home. Although he employed cutters, many his own family, at various times each piece was somehow either chosen, cut, or sold by this man. And certainly most pieces from that era were carefully washed and packed by his wife, Margaret Jane.

Elegant Glass in Corn Flower not only documents the glass used by W.J. Hughes, with its wonderful patterns and colors produced by famous and lesser-known American glass manufacturers, but showcases the beauty and skill of the now famous Corn Flower design.

We can only imagine, through the research and identification by Walter Lemiski for this book, that your favorite piece of Corn Flower will become more treasured as the secrets of its creators, manufacturers, and cutters are revealed. We catch a glimpse of what once was, and is no more, but enjoy the senses and emotions that it brings us now.

Wayne Townsend
Director, Dufferin County Museum & Archives

Dear Readers

While it might be the lot of this author to be ever-so-slightly envious of others who have a wealth of archival materials to work with, such abundance may well take away the wonderful feeling of solving yet another Corn Flower mystery as one delves further into the grail-like search for Corn Flower evidence. So it was recently when we finally realized that the Beaumont Glass Company and Central Glass Works were providers of glass blanks to W.J. Hughes.

This is not a complete Corn Flower pattern reference. Even in restricting the main scope of this volume to the first thirty-five years of Corn Flower, many say the very best years, we took far more pictures than could be shown in one book. As well, there are still undoubtedly many unknowns yet waiting to be rediscovered. In the years since I provided glass identifications and a chapter for Wayne Townsend's book *Corn Flower: Creatively Canadian*, much new information has surfaced. More of the dots have been connected illuminating early production of Hughes's Corn Flower cut. With growing awareness of this splendid Canadian cut pattern, a number of new companies who supplied glass to W.J. Hughes, as well as new information about where original company moulds went after companies dissolved have been discovered. New archival materials have also been unearthed.

In researching approximately twenty different glass companies and hundreds of assorted items, errors are bound to creep in. At times shapes of items made at several different glass shops were virtually identical. Through the study of original advertisements one can usually come to some conclusion as to which company was the likeliest to have supplied W.J. Hughes at a particular time, but some pieces, such as console bowls, pitchers and plates, are at times simply too generic to hazard even an educated guess. The close geographic proximity of a number of glass factories, the transient nature of the work force, the amalgamation of companies, and the sell-off of assets after dissolution of firms have all added to the uncertainty of attribution of a number of items. Scarcely known companies and little documented patterns from even the major manufacturers spice the mix. This project would not have been nearly as thorough even a handful of years ago. The present generation of glass researchers has done much to shed light on many fascinating historical details, and we hope that misattributions are minimal.

We are beginning to more fully realize the importance of the historical ties between W.J. Hughes Corn Flower and most of the major American glass manufacturers of the twentieth century. There are more items yet to be documented and I invite you to join in the ongoing efforts to unravel Corn Flower mysteries. Please feel free to contact me or the Dufferin County Museum and Archives should you have any exciting Corn Flower items, information about blanks, or additional historical details. Who knows what other new and important information may appear in future about the early years of Corn Flower glassware?

Yours in Corn Flower Collecting,
Walt Lemiski
walt@waltztime.com
www.waltztime.com

Acknowledgments

To my sadly missed friend and mentor Edith Hacking, the grand lady of Canadian Depression Glass, past Director of the Canadian Depression Glass Club and promoter of the Glass With Class Shows, who guided me in learning not only about glass, but most importantly about integrity. Edith freely shared her vast knowledge of the field of Depression Glass. She also sagely warned me not to get drawn into the quagmire that is Corn Flower research, knowing that the oh-so-numerous companies Hughes dealt with over so many years would be like a Depression Glass Rubric's Cube. I heeded most of her advice.

To Wayne Townsend, curator of the Dufferin County Museum and Archives, for having the vision to create a world-class collection of Corn Flower and to preserve archival materials related to the collection. A thousand thanks for providing unfettered access to the archives and collections throughout my years of research about Corn Flower. Many of my initial researches were published in Mr. Townsend's book, *Corn Flower: Creatively Canadian,* and in the DCMA's "Corn Flower Chronicle," the newsletter for Corn Flower enthusiasts.

To Barbara and Jim Mauzy for their great friendship, encouragement, and photographic prowess. Also, for their leadership over the past number of years in the field of glass publications — pushing the field to improve with their well thought out innovations. For their valuing and promoting of Canadian glassware, Corn Flower and more, in *Mauzy's Depression Glass,* showing a glimpse of what the Canadian glass industry was creating in the first half of the 20th century.

To Myrna Garrison for sharing her insights into the workings of the Imperial Glass Company and for keeping me on track with her encyclopaedic knowledge of Candlewick. Her meticulous scholarship serves as a model for other glass researchers.

To the super collectors and fine friends who shared their superb collections with me and with legions of fellow collectors: Brenda Beckett, Patrick Doherty, Lynda Edwards, John Lovell, Mike Lutes, Frank and Shirley Martin, Joe Seely, and Brian Wing,

To many friends and dealers who have shared, shown, and alerted us to exceptional Corn Flower items many thanks: Wayne & Jean Boyd, Vi & Ken Brennan, Ken & Teri Farmer, Patti Farrell, Dave & Hilda Proctor, and countless others.

To Douglas Congdon-Martin, my editor, for helping to add clarity and consistency to my jottings.

To the many fellow glass researchers who without reservation have shared knowledge and assisted in verification of facts to make this book as accurate as possible:

Robert Carlson, Elegant Glass
Ed Goshe, Tiffin Glass
Michael Krumme, Paden City Glass
Emily Seate, Fostoria Glass
Tim Schmidt, Central Glass Works
Dean Six, West Virginia Museum of America Glass

Notes on Using this Book

When known, we have utilized the terms from original company catalogues. It is felt that the charm and delight inherent in early variations of terminology would be sorely missed if we tried to codify all item terms. Variations in naming items occur between the numerous companies from whom Hughes purchased glass. Even within a single company's nomenclature, over the years one can find nifty marketing variations for the same item type (e.g. bowl, nappy, bon bon, mint). Items are arranged within sections alphabetically by their most frequently utilized names (e.g. bowl, candlestick, plate). Please look to the Index of Shapes for this ordering.

Since the vast majority of glassware in this volume was hand-made it is common to observe variations of half an inch or more on various items. One of the most often seen variations occurs with bowls. The edges may have been crimped, rolled in, or flattened out. Such is the nature and part of the joy of collecting hand-made glass. Measurements for items with handles, such as bowls and relishes are from the outer edge of each handle, handle to handle. Measurements for pitchers, tumblers, and vases are for their height. Essentially all other measurements indicate length of an item.

About Pricing Corn Flower

Pricing, as anyone in the field can attest, is a tricky, if not treacherous, pastime. Having been involved in buying and selling thousands of Corn Flower items over the past ten years has given me a little insight. Pricing has also been tracked at Depression Glass Shows (probably the best source for finding quality Corn Flower at fair market value from reputable and knowledgeable dealers), general line shows, malls, shops, at live auctions, in the Canadian Depression Glass Review, on Ebay, and through consultation with major dealers and collectors from coast to coast. Values suggested are for items in mint condition. Chips, cracks or other damage will significantly lower a piece's value. The prices give a reflection of the market at the time of writing.

Although both crystal and coloured Corn Flower glassware have long been collected and cherished by collectors, it is the early items that are the most prized. The added details that lower labour costs allowed and the delightful palette of colors from the mid-1920s through to the 1940s, add a further dimension for Corn Flower lovers. Currently all colored Corn Flower items are in great demand. Values for amber, pink, green, and yellow are essentially equal for similar items. Rarer colours, such as amethyst, blue, or vaseline, may well command a premium. Crystal items are generally the most reasonably priced of Corn Flower pieces. Rarer items in the Imperial Candlewick lines however are the exceptions in crystal and will garner amongst the very highest prices of all.

All values in this volume are given in U.S. dollars. Almost all significant glass books are priced in American currency. Web sites, such as the Glass Mega Show, and Ebay auctions have made most Canadian and overseas collectors comfortable in dealing with the U.S. currency. At the time of writing $1 US = $1.35 CDN

This book is intended only to be a guide for determination of an item's value. One must ultimately decide for oneself just how desirable a certain piece is for one's own collection. Neither the author or publishers assume any responsibility for any transactions that may occur because of this book.

Introduction:
W.J. Hughes Corn Flower

The Hughes Corn Flower pattern is distinctive with its 12-petalled flower, grid-like interior, and elegant sweeping stems. Corn Flower glass was produced for over half a century. The most sought after items are the pre-1950 items. First among these are the colored items, followed by the crystal (clear) items with excellent cuts, and the ever-popular Imperial Candlewick. One can recognize the earlier items by their fuller cutting, larger flowers, and a tendency towards a wreathing effect with the leaves surrounding the floral pattern. Many earlier blanks also display "beading" – a series of cuts, like a piecrust, found at the edges, handles, and bases of pieces. On smaller cutting surfaces one might also find a "bud" variant – a round portion with three fringed petals. The bud was not in use much after the early 1950s and beading was totally phased out around 1953. The added details that lower labor costs allowed and the delightful palette of colors from the mid-1920s through to the 1940s, add a further dimension for Corn Flower lovers.

Heisey #1883 "Revere" Bonbon, 2-handled, 8". *Courtesy DCMA.*

Tiffin #336-2 Cake Plate, handled, 11", $200. Note small version bud cut around rim; piecrust edging, called "beading" by W.J. Hughes, around the edge. *Courtesy DCMA.*

The opening shot from the 8 mm film circa 1948 about W.J. Hughes creating his trademark Corn Flower pattern. In front of the Corn Flower sign are an assortment of Candlewick items and stemware.

The founder of the W.J. Hughes Corn Flower Company, William John Hughes (1881-1951), was born in Dufferin County, Ontario, Canada. As a young man, around 1900, he was employed by Roden Brothers silversmiths of Toronto. Fortunately for him and for glass collectors, when that firm expanded their lines to include cut lead glass Hughes was asked to learn the art of glass cutting. Hughes family oral history suggests that, starting in 1912, he began to experiment at his home with his own original "grey" cut glass patterns (shallower cuts on lighter glass). As of yet there have emerged no examples of Corn Flower cut glassware that can be definitively dated any earlier than 1925. So far, any of the identified glass blanks that were in production early enough to have been cut in the teens by W.J. Hughes were still being produced decades later. For example, A.H. Heisey's #1184 Yeoman line was made from 1913 through to 1957. Dating of any items to within the first ten years of Corn Flower cutting must be purely by conjecture. Existing documentation indicates that Corn Flower started to be commercially cut in either 1914, the starting date as found in the 1940 Corn Flower word trademark papers, or in 1916, the starting date stated in the 1949 Hughes vs. Sherriff law suit (see Appendix D for more about R.G. Sherriff).

Close up of W.J. Hughes cutting Corn Flower, 8 mm film, circa 1948.

W.J. Hughes in his shop at the cutting wheel. From a still made from the 8 mm film showing the master cutter creating his Corn Flower creations, circa 1948.

Pair of Heisey Saucer Champagnes, photographed at a quarter turn showing both a Corn Flower and a standard bud cut on the same stem, c.1925, $25 each. *Courtesy DCMA.*

Through the first several decades of Hughes Corn Flower production, the blanks used for cutting came primarily from the tri-state area of Pennsylvania, Ohio, and West Virginia. These three states were home to all the major American glass producers and accounted for over ninety percent of glass production in North America throughout this period. Fine quality glassware was essential for the Corn Flower decoration that Hughes cut. What one thinks of as typical Depression Glass is scarcely better than bottle glass. Inherent in such low-grade glass were bubbles, lack of clarity, and non-uniformity of shape. What Corn Flower was primarily cut upon was "Elegant Glass," also known as "hand glass". Most of the companies that supplied glass to Hughes were major players in the American glass field including Cambridge, Duncan Miller, Fostoria, Heisey, Imperial, Indiana, Jeannette, Lancaster, New Martinsville, Paden City, Tiffin, Viking, and West Virginia Specialty glass companies.

North American glass manufacturing underwent major changes between the two World Wars. During the first decade that W.J. Hughes began acquiring glass blanks (1916-1925), American glass producers were essentially manufacturing what collectors today refer to as "crystal" glass. In Depression Glass collectors' lingo,

this simply means clear uncolored glass and does not refer to lead crystal. No Canadian glass companies were producing the quality of glass that Hughes required for his decorating.

Starting in the mid-1920s, coinciding with the impact of mass-production on the glassware marketplace, colored glassware began to emerge as an important product. With the diminishing costs of production, thanks to new technologies, glass companies could afford the little bit extra required to color the glassware. Company after company quickly jumped on this color bandwagon. The basic range of colors was transparent amber, green, pink, and yellow, although the spectrum of glass colors was widened with various shades of these basics and through experimentation with opaque glass. The September, 1929, issue of the influential homemaker's magazine *Better Homes and Gardens* included an article entitled "Serve It in Colored Glassware." The author declared that: "The aristocracy of colored glassware is in its richness and warmness, and the former preponderance of the frigid and sheer-crystal glassware on our table wholly lacked this. Whatever it is, serve it in glass!"

11

Heisey #1229 "Octagon" Flamingo (pink) Hors' D'Oeuvre Plate, 13", $225; Heisey #1229 "Octagon" Flamingo Dessert Dish, Oval, 8", $225. The "Octagon" pattern was made by Heisey between 1925 and 1937. *Courtesy DCMA.*

Although one tends to think automatically of Depression Era glassware in terms of its distinctive colors, keep in mind that crystal was never totally gone from the marketplace. From mid-1920s through the 1940s, colors were available and utilized for Corn Flower. However, large quantities of the tried and true traditional crystal wares were still being produced throughout the 1930s. By the mid-1940s, fashion had shifted once again, and color was virtually gone from the marketplace. Besides a very small number of ruby-footed and handled items from the late-1940s, almost all colored glassware cut with Corn Flower was produced between 1925 and 1945. The couple of exceptions presently known are from the Viking Glass Company: a large amethyst bowl, and a ruby covered candy jar produced in the early 1960s. The greatest run of colored Corn Flower would fall between 1926 and 1940.

New Martinsville #103 Candy Box, 6-1/2", $300. *Courtesy DCMA.*

New Martinsville #37 "Moondrops" 4 oz Tumbler, 3-1/2", $50; #37 "Moondrops" Cocktail Shaker, 9-1/2", $225. *Courtesy DCMA.*

Lancaster #R767/6 Topaz Rose Bowl, 6", $155. *Courtesy DCMA.*

What was essential for Hughes, first and foremost, was quality of glass. The mass-production of what we now call Depression Glass involved huge tanks of glass that were fed into machinery to produce massive quantities of glassware. The whole process involved very little quality control and virtually no hand finishing. Inherent in such low-grade glass were bubbles, lack of clarity, and non-uniformity of shape. To cover such deficiencies, many Depression Glass lines were molded in patterns that covered much of the glass surface, thus disguising the irregularities. During the same era, a superior grade of glassware was being produced known today as "Elegant Glass."

Elegant Glass was produced in considerably smaller batches with a better quality of raw materials (the glass metal). During production this glassware often underwent further stages with more finishing or decorating done by teams of skilled glass craftsmen and women. Gold trims, ground bases, cuttings, plate etchings and fire polishing were among the extra measures taken with this finer quality product. Hughes required blanks that not only were of superior quality, but were uncluttered in design and left large clear surfaces on which to better showcase his Corn Flower design.

Tiffin #020 Saucer Champagne, 4-3/4", with gold encrusted band, $45. *Courtesy DCMA.*

Pair of Imperial Candlewick #400/74SC Bowls, crimped, 4-toed, ribbed. Note variation in crimping, $75 each. *Courtesy of Brenda Beckett.*

To give some notion of the scope of glass production in this era, we only have to look to West Virginia in the late 1930s. Given years of downsizing, amalgamations, and the Depression, there were still over one hundred and fifty glass factories in 1938 in West Virginia alone. This number included pressed tableware producers and Elegant Glass factories, as well as mills manufacturing everything from bottles to windshields. Add decorating shops to this group and number of craftsmen involved in the glass industry was quite substantial.

Some twenty different sources of glass blanks have been identified from this period as suppliers to W.J. Hughes during his thirty-five years in business. There is a distinct lack of documentation about Hughes Corn Flower prior to 1950. Many of the company's records were destroyed as recently as the 1980s upon dissolution of the firm. Also, W.J. Hughes apparently believed that word-of-mouth was the best way to grow a business. Searches of business directories and phone books from this era fail to turn up any listings for W.J. Hughes as a cutter of glassware. Consequently, and unfortunately for researchers, he also did virtually nothing to advertise his wares through standard trade magazines or print media, usually an excellent source of information. Inquiries made to glass museums and associations concerned with the companies that Hughes dealt with have also proved fruitless.

What we have to work with consists of notations made from a few mid-1920s invoices, the 1938 Thornley Wrench catalogue, a grouping of bank draft papers, primarily from the 1940s, the decision from the 1949 Hughes vs. Sherriff law suit, and the wonderful existing legacy that is the cut glass of W.J. Hughes.

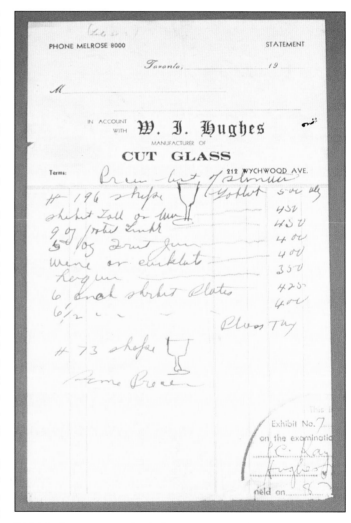

W.J. Hughes invoice with wholesale price list of stemware, c.1937.

Part I

Early Corn Flower Blanks: The Glass Providers

Beaumont Glass Company (1890-1993)

Percy J. Beaumont, an Englishman, set up the Beaumont Glass Company in Martin's Ferry, Ohio in 1890. His brother-in-law was none other than the glass-making great Harry Northwood. Many published details of this firm's beginnings and products remain murky and contradictory. However, after various relocations the glass works settled in at Morgantown, West Virginia, around 1915. It was at this time that Beaumont created the glass it is most recognized for today, Fer-Lux, a clambroth-coloured glass. Fer-Lux was used primarily for light globes and lamp shades, and to a lesser extent for tableware and vases. Frequently, Beaumont pieces were adorned with dot or floral motifs, done in their own decorating shops. Presently a number of trays and vases have been identified as items cut by W.J. Hughes with his Corn Flower design. Pictured in Hughes's 1938 catalog is an entire page featuring six Beaumont 3-footed rose bowls in a variety of forms (see page 147). Beaumont Glass closed in 1993. There is undoubtedly much yet to be discovered about this company's century of glass making.

Cambridge Glass Company (1901-1954)

The Cambridge Glass Company was actually chartered in 1873. However, it was not until 1899 that land was purchased for the factory. This prestigious firm was initially owned by the National Glass Company of Pennsylvania. They hired the British-born Arthur J. Bennett to manage the plant. Bennett designed the very first item produced in May, 1902, a crystal 3-pint jug. By 1907, National was having severe money troubles and the company was put into receivership. Bennett leveraged enough money, some $500,000, to buy Cambridge Glass outright. In the early years Cambridge had their own coal mines from which they used fifty tons daily. Producing only hand glass, no "automatic" glass, the firm was thriving through the 1920s. At peak periods they had a work force of 700, ran three shifts per day, and used a staggering fifty-six pots of glass. It was also in the twenties that they started colored wares, complete dinner services, and their use of their trademark "C" in a triangle.

Beaumont Rose Bowl, crystal, 3-footed with flower frog, 3-1/2", $65. *Courtesy DCMA.*

Cambridge #156 "Martha" Oyster Plate, crystal, 6-part, 10", $75.

Cambridge produced everything from figurals to pharmaceutical items. They produced some five thousand moulds over their half century of manufacturing and were known for the quality of their glassware and for their rich color palette. Among their best known products were the 3400 line, Caprice, Statuesque, and the Rose Point etching. Fairly little Cambridge glass has been found cut with Hughes Corn Flower. Amongst the items pictures in this volume are the #933 amber cup and saucer, the #156 Oyster Plate, and a series of vases seen in the 1938 Thornley Wrench Corn Flower advertising photographs.

Central Glass Works Bowl, amber tri-cornered, 3-footed, 6-1/4", $200. *Courtesy DCMA.*

Central Glass Works (1896-1939)

Although Central Glass Works roots go back as far as 1863 and the company was one of the biggest and best of Depression Era firms, astonishingly little information has emerged until recently. In August of 1891 the original Central Glass Company of Wheeling, West Virginia, was amalgamated into the large U.S. Glass Company conglomerate. It was known as Factory "O". Another big player in Corn Flower history and fellow USGC member, was the Tiffin Glass Company, Factory "R". In 1896, Central's moulds were shipped off to other factories of the behemoth. It was at this time that locals in Wheeling purchased the factory and it became Central Glass Works.

By 1924, Central Glass Works had a large contingent of nearly two hundred-fifty workers. The *National Glass Budget* of January 15, 1927, stated that: "Central Glass Works makes a full line of high-grade table ware and novelties, decorated and gold encrusted blown stemware. It, also, is showing pressed stemware in colors and Chippendale tableware. This company claims to be the pioneer in manufacturing the popular rose colored glassware, as well as the originators of the salad plates. It is also showing a beautiful line of exquisite shaped, copper wheel engraved high goblets and low English shaped goblets in rock crystal cutting. Its new orchid colored glass presents a beautiful appearance." Central Glass Work's emphasis on quality goods was reflected in their ad motto "The Old Central Quality". They maintained a significant business in supplying blanks to decorating companies, like W.J. Hughes. Central declared bankruptcy in 1938. Many of their moulds were purchased by the Imperial Glass Company in 1940.

Unfortunately, a number of Central Glass Work items are easy to confuse with other competitors' products. Their line of serving pieces, #2019, is quite close in style to Cambridge lines, but not exact. That is not the main problem. New Martinsville Glass also produced a #2019 line whose items are identical to the Central items. Whose is whose? Similar conundrums arise between Central and Paden City Glass items.

The "Deco Fan" line of items was among the moulds that Imperial reused. Starting around 1943, Imperial ran this line for use with their #148 Belmont Hills cutting especially for Bechtel, Lutz, and Jost Company of Reading, Pennsylvania. With Hughes's close association with Imperial and the amount of this blank found with Corn Flower cuttings, it is thought that these blanks most likely came from Imperial and not from Central. As further research is conducted hopefully some of these Central mysteries will be solved.

Central-Imperial #1480 Salad Bowl, crystal, 2-handled, 10", $55; #1480 Cake Plate, 2-handled, 11", $50.

Duncan Miller Glass Company (1893-1955)

The Duncan Miller Glass Company enjoyed a very long and successful enterprise. Initially formed in 1865, production began in earnest in the 1870s in Pittsburgh, Pennsylvania, as George Duncan & Sons. Like Central Glass Works, the company merged with the group of companies that were together known as the United States Glass Company, the firm out of which the flagship Tiffin

Glass factory later emerged in the 1920s. In 1892, the factory was destroyed by fire. At that point the owners left the glass consortium and opened a new factory in Washington, Pennsylvania. By the turn of the century, it became known as the Duncan Miller Glass Company (named for its principals Harry Duncan, James Duncan, and John Miller.) Amongst their most celebrated lines were Canterbury, Sandwich, Hobnail, and Teardrop.

The quality of their materials and their care and craftsmanship placed Duncan Miller firmly in amongst other Elegant Glass producers. They were well known for their use of vibrant colours: green, pink, cobalt blue, light blue, black, and ruby. The most readily recognized Duncan Miller items cut with Corn Flower are the Pall Mall No.30 pattern swans. An excerpt from a Duncan Miller catalog circa 1945 states: "Pall Mall, the 'modern' style. As such, it lends itself to a myriad of new uses — for home decoration, for the table, for flowers."

A number of Duncan Miller relishes appear in the 1938 Thornley Wrench catalog. In the mid-1950s, the U.S. Glass Company bought out Duncan Miller. Some of these moulds were used at their Glassport, Pennsylvania factory (one of the Tiffin factories.) The Fenton Art Glass Company acquired some of the Duncan moulds in the 1960s.

European Glass Companies

Although many particulars remain obscured, we do know that W.J. Hughes purchased some pre-1950 glass blanks from firms in Czechoslovakia and Poland. The largest number of Corn Flower cut Candlewick look-alikes came from Czechoslovakia. They have alternately been called European or Czech Boule. Evidently, R. Schrotter designed these Candlewick-like items in 1935 for the Rudolfova Hut at Inwald, Czechoslovakia. These items were manufactured by the state run Glass Union. Unfortunately, once again, no invoices have emerged. However, a single bank draft for $1316 with the cryptic notation "re L/C 3770" still exists indicating that money was transferred to Prague on May 10, 1948.

Czechoslovakian "Boule" Vase, crystal, 9-1/2", $85. *Courtesy DCMA.*

Duncan Miller #30 "Pall Mall" Swans, crystal, 5", $35; 7", $45; 11", $95. *Courtesy DCMA.*

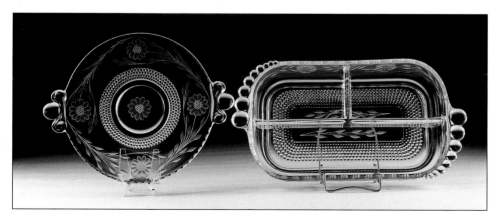

Duncan Miller Plate, 2-handled, 10", $35; #301 "Tear Drop" Relish, crystal, 3-part, 2-handled, 10", $45. *Courtesy DCMA.*

17

Czechoslovakian "Boule" Bowl, amber,
3-footed, 8", $200. *Courtesy DCMA.*

Federal "Park Avenue" Vegetable Bowl, crystal, 8-1/2", $25.

Federal Glass Company (1900-1984)

The Federal Glass Company formed in Columbus, Ohio, in 1900. In the early years the company was a hand-glass operation. They had various shops to add decorations and needle etchings. Federal was fast off the mark when the automation revolution occurred in the glass business in the 1920s. Among its best-known Depression Era lines were Madrid, Parrot, and Sharon. It was in this period that they became one of the largest producers of automatically produced tumblers and pitchers. In 1958, they were incorporated with the Federal Paper Board Company. The much higher profitability of the paper division led to Federal Glass Company's closure in 1984.

Federal Glass was the supplier of choice of W.J. Hughes for flat tumblers from the 1940s onwards. Their heavy base sham tumblers were bought in such large quantity over so many years that they are very readily found to this day. When Federal announced that they were closing up W.J. Hughes Corn Flower purchased two trailers of their tumblers. These very tumblers were the last items ever to be cut with Corn Flower.

Fostoria Glass Company (1887-1986)

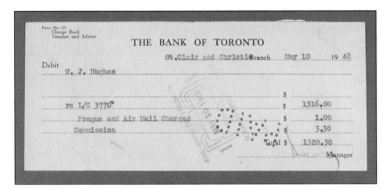

Bank Voucher, 10 May 1948, Prague. *Courtesy DCMA.*

The Polish connection is more nebulous still. In the mid to late 1940s some ruby-footed pilsner glasses and some ruby-handled beer mugs were ordered from a glass factory in Poland. The lengthy, uncertain delivery time discouraged repeat business with this company. A second, and last, order was received in 1949.

Polish Beer Mugs, crystal with red handles, 4", $60 each. *Courtesy of Brenda Beckett.*

Fostoria #2496 "Baroque" Relish, crystal, 3-part, 12-1/4", $35. *Courtesy of Brenda Beckett.*

The Fostoria Glass Company began in Fostoria, Ohio, in 1887. They remained at that location until the source of cheap natural glass that enticed them there began to run low, whereupon they relocated to Moundsville, West Virginia in 1891. Fostoria started out producing oil lamps, pressed glass for home and hotel, vases, and candelabra. By the early 1900s, Fostoria changed focus to glass for the household – from stemware to tableware and occasional pieces. They introduced their durable cubist American pattern as well as many more refined lines with a large variety of etchings. By the 1920s, they ranked as one of the foremost of the hand-glass producers. In 1924, Fostoria introduced colors and, by 1926, they brought out the first complete glass dinner services. Various Fostoria lines appeared in no fewer than a dozen different colors, as well as in crystal. One color for which they were noted was Topaz, a golden yellow shade. They had a chemical research laboratory and a special design department replete with top artists. They produced quality products and, naturally, their

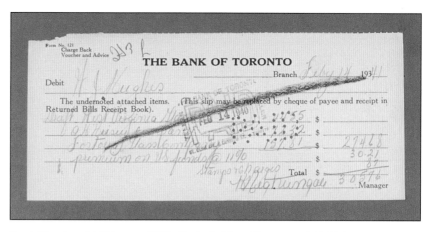

Bank Voucher, 14 February 1941, Fostoria Glass Company, A.H. Heisey Glass Company, and West Virginia Specialty Glass.

glassware was priced accordingly. Perhaps this is the reason that not a lot of Fostoria glassware can be found cut with Hughes Corn Flower – the price was simply prohibitive. Hughes aimed his marketing, such as it was, at the middleclass consumer. Items cut on Fostoria Baroque, Coronette, and Lafayette patterns are occasionally to be found. The Fostoria fires were finally cooled in 1986 after nearly a century of extraordinary production.

A.H. Heisey Glass Company (1893-1956)

Augustus H. Heisey opened his glass factory in Newark, Ohio, in 1896. Through its sixty years of production, Heisey Glass became synonymous with quality glassware. Heisey's renown arose from the purity and consistency of basic materials. A fair number, but not all, of Heisey's items are signed with their trademark "H" in a diamond. W.J. Hughes ordered from several of Heisey's major glass lines: #1184 Yoeman, #1229 Octagon, and #1483 Stanhope. He also bought a wide variety of pieces including baskets, bonbons, candlesticks, celeries, colognes, flower bowls, relishes, and other items.

Some of the scant paper evidence from prewar invoices of glass purchases by W.J. Hughes comes from secondary sources. Early in 1969, Pete Kayser and Robert Prouse of Hughes Corn Flower corresponded with Lucile J. Kennedy, Assistant to the President of the Imperial Glass Corporation, and with the Heisey researcher Clarence W. Vogel. Vogel was in the midst of compiling his four volumes on Heisey glassware. The letters mention invoices dating from the late 1920s. From these

notes we learn more about the scope of blanks utilized. Unfortunately, no information about the size of these orders was recorded. The only indication of the color of these items was for the #113 Mars candlestick that was labeled "Flamingo," the Heisey term for pink, their top selling color produced between 1925 and 1935. We have spotted various of these items in crystal, Moongleam (Heisey's green), Sahara (Heisey's yellow), and Hawthorne (their pale amethyst).

Heisey #1495 "Fern" Whipped Cream or Mayonnaise, crystal, 5", $40; Plate, crystal, 6", $30; Mint, crystal, handled, 6", $35. *Courtesy DCMA.*

Item No.	Line Name	Production Dates	Item Description	Invoice Date
112	Mercury	1926-1957	Candlestick 3"	23 Sept. 1926
113	Mars	1926-1927	Candlestick 3"	23 Sept.1926
1023	Yeoman	1922-1937	Creamer & Covered Sugar	25 May 1926
1183	Revere	1913-1935	Jelly, 5-1/2", 2 handled	
1185	Yeoman	1922-1937	Celery Tray, 12"	
1186	Yeoman	1913-1957	Cup and Saucer	
1229	Octagon	1925-1937	Cheese Dish, 6", 2-handled	
1229	Octagon	1925-1937	Mint, 6", 2 handled	
1229	Octagon	1925-1937	Muffin Plate, 10", 2-handled	
1229	Octagon	1925-1937	Muffin Plate, 12", 2-handled	22 Oct. 1926
1229	Octagon	1925-1937	Sandwich Plate, 10", 2-handled	
1229	Octagon	1925-1937	Sandwich Plate, 12", 2-handled	22 Oct. 1926
3350	Wabash	1922-1939	Comport and Cover, footed, 6"	23 Sept. 1926
4163	Whaley	1919-1953	Tankard, 54 oz	

—Notations from the 1926 Invoices

Heisey #3350 "Wabash" Comport and Cover, Flamingo (pink) 6", $300. *Courtesy DCMA.*

Heisey Glass Company was one of the major suppliers of glass for Corn Flower through the early years. After having been at the forefront of the glass industry for some sixty years, the Heisey glass works closed in 1956, at which time the Imperial Glass Company purchased their moulds.

Imperial Glass Company (1904-1984)

Imperial Candlewick 400/156 Covered Mustard with Spoon, crystal, $65; 400/56 Relish, crystal 5-part pinwheel sections, 10-1/2", $60; 400/259, Candy, crystal covered, shallow round, 7", $120. 400/119 Oil or Vinegar with Stopper, crystal, $55.

Other early pieces that W.J. Hughes ordered from Imperial included items from the Pie Crust line, the Crochetted Crystal line, and, one of my favorites, No.100, the boot toothpick holder. By the mid-1950s, W.J. Hughes Corn Flower became the sole agent for Imperial Glass in Canada. Imperial's relationship was long and fruitful with Hughes Corn Flower.

Bank voucher, 3 March 1947, Imperial Glass Company. *Courtesy DCMA.*

The Imperial Glass Company began its long tenure at Bellaire, Ohio, in 1904. Producing glass until 1984, Imperial was one of America's premiere hand-glass (Elegant Glass) manufacturers. Apparently, Hughes's long association with the Imperial Glass Company began in the late 1930s. Although not represented in the 1938 Corn Flower advertising photographs, Imperial Candlewick cut with Corn Flower would emerge at the end of the company's third decade as the most enduring of all their lines (see Part IV for details about Corn Flower on Candlewick).

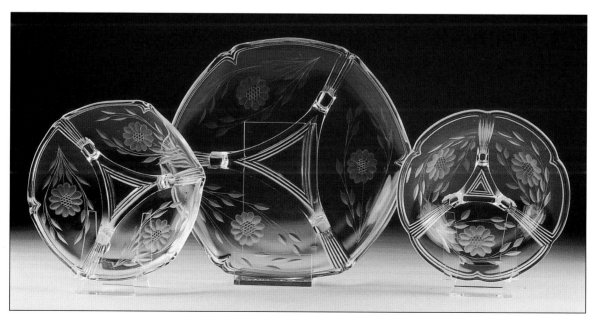

Imperial #5885B "Pie Crust" Bonbon, 3-footed, 6-1/4, $25; #5889F, Bowl, 3-footed, 9-1/4", $35; #5886C, Tray, 3-toed, 6-1/2", $25. *Courtesy DCMA.*

Indiana Glass Company (1907-2002)

The Indiana Glass Company is well known to Depression Glass collectors for such standard patterns as Avocado, Pretzel, Pyramid, and Tea Room. Indiana Glass started business in Dunkirk, Indiana, in 1907. They emerged out of the Ohio Flint Glass Company and the National Glass Company. One of their ads stated that they were "Manufacturers of pressed and blown glassware, hand-made and machine-made complete lines of tableware, bar goods, hotel ware and soda fountain supplier." They also produced decorated ware. This is of note for Corn Flower glass research, since a number of ruby flashed plates and candlesticks have been located cut with Corn Flower. Among the Indiana patterns often seen cut with Corn Flower from the 1940s is #607, a pattern featuring a fleur de lys motif, and #1011, "Tear Drop".

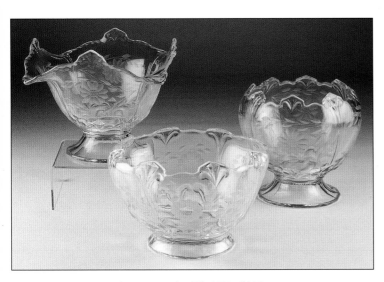

Indiana #607 Comport, 10-1/2", $55; #607 crystal Comport "E", 8", $55; #607 crystal Comport, 7", $55. *Courtesy DCMA.*

Jeannette Glass Company (1898-1975)

The Jeannette Glass Company began production in Jeannette, Pennsylvania, in 1898. In its early days the company produced such items as "vault lights, prism tile, packers' ware, and novelties." During 1927-28, brand new machinery was installed, with an automatic glassmaking system that could mass-produce some fifty tons of glass daily with two continuous tanks. It was at this juncture that the Jeannette Glass Company ventured away from hand-finished glassware. Unlike most of the other firms from which Hughes purchased glass in the 1920s and 1930s, who were all Elegant Glass companies, the Jeannette Glass Company is known as one of

the major players in production of what is today called Depression Glass, that lower cost, lower quality mass-produced glassware that became possible in the mid-1920s due to revolutionary new technologies. Amongst other innovations, Jeannette even suggested in their advertising that they may very well have been the first company to produce pink and green glassware automatically in a continuous tank.

The Jeannette line number 5186 items, referred to in notes taken from Corn Flower invoices dating from 1927, are bell-shaped bowls with black bases. The colors noted were amber, crystal, green, and topaz (yellow). None of these has surfaced recently, but one suspects that only the bowls would have been adorned with Corn Flower cuttings. What are found occasionally are the #102 green sherbet and salad plates with honeycomb borders. The #82 nappies mentioned in the invoice have yet to be uncovered.

Jeannette #102 Salad Plate, 7". $55. *Courtesy of Brian J. Wing.*

From the February 20, 1928 issue of the *China, Glass and Lamps* trade paper, we have the following notice that may well have intrigued W.J. Hughes:

"Another item of general interest was the machine-made salad plate. There were three designs shown, including one plain salad plate for use in the decorating trade. Being automatically made, these plates are uniform and can be stacked to a height of six feet or more if necessary. This is a great advantage to the decorating trade and also to the department store buyers, because lack of space in various glass departments make it necessary to stack plates and if they are not uniform it not only takes more room but also shows the irregularity of the plate as soon as the customer sees the stack."

Previous theories about Hughes only purchasing fine quality Elegant Glass have been dashed. However, very little Jeannette glassware has emerged to date, leading one to suspect that their associaton with Corn Flower may well have been a very short-lived one. Certainly the crystal clear, fine quality that one expects to see in the wares that Hughes generally cut is not to be found in Jeannette glassware of this era.

Lancaster Glass Company (1908-1937)

The Lancaster Glass Company set up shop in Lancaster, Ohio, in 1908. The firm was run by two brothers, Lucien and Philip Martin, who had worked for the Hocking Glass Company. By 1924, Lancaster had been amalgamated into the larger Hocking company. Besides tableware, Lancaster produced a large array of occasional items such as bonbons, bowls, and vases. Their in-house work on decorating puts them in the category of Elegant Glass houses. The quality of their glass, however, was on the lower rungs of the Elegant lineage. While the quality was quite good, it certainly was not up to the high standards of the Cambridge, Fostoria, or Heisey Glass companies. Although the factory remained in use for many more decades, the Lancaster name was in use only until 1937. From that date on the parent company, the Hocking Glass Company, began to market all their subsidiary's products under the Hocking banner.

Lancaster colors of interest to Corn Flower collectors include green, produced from 1925, deep pink introduced in 1926, and topaz (the trade name for yellow) made from 1930. Items such as the center-handled sandwich tray, Lancaster No.88, appeared in 1923 and were specifically intended for firms like Hughes Corn Flower, as the ads stated they would be ideal for "light cutters and decorators." Topaz is the most prominent Lancaster color found with Corn Flower cuttings. Most Lancaster-Corn Flower items are distinctive with their pointed edges, ornate handles, and scroll feet. Candy dishes, rose bowls, center-handled sandwich trays, two-handled cake plates, and candlesticks are among items that were utilized by Hughes.

Lancaster T1831-7 pink Tray, 8", $125; T1831-7 Tray, 10", $155. *Courtesy DCMA.*

Mid-Atlantic Glass Company (1937-present)

As happenstance would have it, due to a work stoppage at the Louie Glass factory in Weston, West Virginia, in 1937 a group of eleven craftsmen decided to set up their own glass works. Mid-Atlantic Glass was established in Ellenboro, West Virginia. As joint owners they worked at what they knew best – making mouth blown and hand glass. Incredibly their staff has numbered steadily around one hundred workers for over half a century. Mid-Atlantic is a little known firm since they sell directly to distributors, wholesalers, and, importantly for Corn Flower, to decorating shops. Much Mid-Atlantic glassware is very difficult to attribute since they continued on producing items very similar to those made at Louie Glass. Among their most distinctive and collectible items are those with colored feet and handles in green and red. These colored wares were produced between 1937 and the early 1950s. Items most frequently seen and recognized with the Corn Flower cut are the colored-base pilseners.

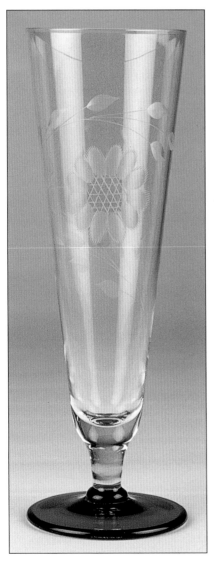

Mid-Atlantic, Pilsener, crystal with green foot, 8", $55. *Courtesy of Mike Lutes.*

New Martinsville Glass Company (1900-1944)
Viking Glass Company (1944-1970)

Bank voucher, 7 July 1939, New Martinsville Glass Company. *Courtesy DCMA.*

Beginning in the 1930s, the New Martinsville Glass Company became one of the larger suppliers of glassware to Hughes Corn Flower. At their opening in 1900, the New Martinsville plant began to manufacture tableware, lamps, and novelties. In a 1923 edition of the glass industry trade paper, *American Flint,* it was reported that New Martinsville was employing "eleven cutters working full time and cannot keep up with the orders." Aware of the needs of their own in-house cutters, the company was designing blanks with space for cutting in mind. Their own ads stated that "any style of cutting desired" could be supplied.

New Martinsville #37/3 "Moondrops" Candelabra, amber 3-lite, pair $300; #37 "Moondrops" Bowl, ruffled, amber 3-footed, 9-1/2", $250. *Courtesy DCMA.*

Two quite outstanding lines of New Martinsville Company glassware that appear in the 1938 Hughes Corn Flower advertising photographs are Moondrops and Radiance. Moondrops, Line No.37, was introduced at the end of 1932, and features a raised circle on the bottom sections of the blanks "reminiscent of the age of chivalry." It grew to be a large line with over ninety different pieces being made. In the 1938 Corn Flower photographs, a "Moondrops" cocktail shaker with Corn Flower cut is pictured. It is essentially a large mug topped off with a metal lid. Matching whiskey glasses and a cream and sugar set are shown in the same photograph (see page 150). An ad from 1934 in the *Pottery, Glass and Brass Salesman* lists the cocktail set as "The Butler's Delight."

New Martinsville #42 "Radiance" Nut Dish, blue, 5", $175 each. *Courtesy of John Lovell.*

The other major New Martinsville pattern for Corn Flower cutting was line No.42, Radiance. Unveiled in 1936, Radiance, called "Tear Drop" by the Hughes, became one of their favorite blanks and was utilized over many years. New Martinsville's introductory advertisement stated that "while each piece is highly decorative in itself it is so designed as to allow ample space for decorations and cuttings." Colors noted on the 1938 Corn Flower advertising photos indicate that the eleven inch plate may have been purchased in crystal, amber, and blue. (The only known ruby 11" plate is pictured on page 67). For the two-part, eight inch Radiance relish the colors listed were crystal, amber, blue, rose, and green. The Viking Glass Company continued producing Radiance primarily in crystal after taking over the company in 1944. The line was quite an extensive one, numbering some fifty pieces. New Martinsville remained an important supplier for W.J. Hughes under the Viking banner for many years.

Viking #4457, Candlesticks, crystal 2-lite , pair $75. *Courtesy DCMA.*

25

Paden City Glass Manufacturing Company (1916-1951)

Paden City #411 "Mrs. B" Cream & Sugar, pink, 3-1/8", pair $200. *Courtesy DCMA.*

David Fisher, who had been a part of the New Martinsville Glass Company, established a glass plant in Paden City, West Virginia, in 1916. Production was essentially of crystal pressed table wares. Paden City soon grew to be known for its development of over two dozen colors through the 1920s. As with most other of Hughes's suppliers, Paden City was a hand-glass or Elegant Glass producer that made fine-blown tumblers and stemware and had specialized decorating departments. Because of close proximity to New Martinsville, and with the transient work force switching back and forth between the companies, both firms' products exhibit many stylistic similarities.

Paden City Bonbon, 2-handled, 7-1/2", $30; Gravy Bowl, 2-spout, 7", $55; Relish, 4-part, 2-handled, $45; #215 "Glades" Bowl, crystal, 2-handled, 5-3/4", $30. *Courtesy DCMA.*

The 1938 Corn Flower advertising photographs show a full page of Paden City's "Crow's Foot" line (see page 147). Hughes evidently bought a goodly supply of relishes and other serving pieces from Paden City. Major lines to be found with Corn Flower cuttings are: #215 Glades, #220 Largo, #555 Gazebo, and #777 Comet. The Paden City Glass Manufacturing Company remained in business until 1951.

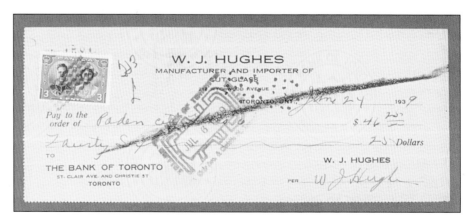

Check, 24 June 1939, from W.J. Hughes to Paden City Glass.

Tiffin Glass Company

The Tiffin Glass Company was a favorite supplier for W.J. Hughes through the 1930s. This is not surprising when one ascertains that during this period Tiffin produced as much as, if not more than, most of the other large glass firms. No less than a quarter of all the items illustrated on the pages of the early Hughes photographs were cut on Tiffin blanks.

Tiffin began as the A.J. Beatty and Sons Glass Company, of Steubenville, Ohio. They moved to Tiffin, Ohio, in July, 1888, thanks to that town's generous enticement of five years of free natural gas, thirty-five thousand dollars, and land worth fifteen thousand dollars. In 1891 it was one of the companies in an association of nineteen glass factories that joined forces as the U.S. Glass Company. The amalgamation of so many glass firms gave U.S. Glass a distinct advantage. Through specialization, sheer size, and use of non-union workers the conglomerate strove to swamp the competition. In 1923, U.S. Glass provided a staggering "25,000 pieces to Choose From!" in its catalogs. Initially, Factory R, the Tiffin factory, specialized in producing hundreds of thousands of pressed glass tumblers per week, along with barware.

Starting in 1914, the Tiffin factory was switched to producing higher quality blown and molded tableware and stemware entirely. By 1927, the brain trust at the United States Glass Company decided it would be advantageous for marketing purposes if their products were known for their quality – the Tiffin quality. The Tiffin factory became the flagship of the organization. When the company sought to cash in on the Tiffin name, it created a separate label for Tiffin – TIFFIN superimposed over a large "T" within a shield on a gold paper label.

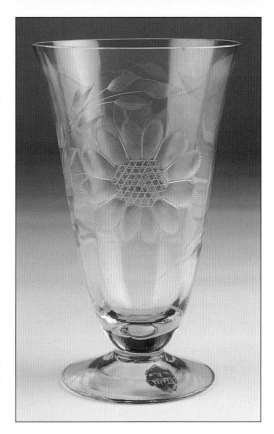

Tiffin #020 Footed Ice Tea, pink, 5-3/4". Note original Tiffin Glass Company label on foot, $65.

In this era, Corn Flower was typically cut on three sizes of footed tumblers: the smallest was the oyster cocktail or whiskey; the mid-size was the seltzer; and the large size was the table tumbler. A half dozen sizes of stems were regularly cut including cocktails, sundaes, saucer champagnes, wines, sherries, and goblets. Many of the Tiffin stemware lines are quite distinctive with their ornate stems. Other Tiffin items cut with Corn Flower include baskets, candlesticks, candy dishes, cake plates, comports, cream and sugar sets, pitchers, plates, relish dishes, trays, and vases.

As with other major glass producers, starting in the mid-1920s the Tiffin Glass Company created a large palette of glass colors. The marketing geniuses were not

Tiffin #15360 Candlestick, green 3-lite, 6-1/2", $325. *Courtesy of Ken and Terri Farmer.*

27

content to call their products simply pink, green, or yellow, but used a delightful array of descriptors: Amberina, Amethyst, Canary, Emerald Green, Lilac, Mandarin, Old Gold, Reflex Green, Rose Pink, Royal Blue, Sky Blue, and Twilite. The colors that tend to be seen most often are the four standard Depression Era colors of Rose Pink, Reflex Green, Canary (yellow) and Mandarin (amber). The stunning Royal Blue and Lilac, scarcer Tiffin colors, are found more infrequently.

Tiffin continued to produce glass through until 1980, although the company had been purchased by a couple of new owners after the 1962 bankruptcy of the original firm. As is the case with other elegant glass companies, as costs rose after 1945, Hughes could not profitably continue to purchase from firms who had been his earlier major suppliers.

Westmoreland Glass Company (1889-1984)

The Westmoreland Glass Company had a fabulous run of almost one hundred years. Rich reserves of natural gas drew manufacturers to Westmoreland County, Pennsylvania, and Westmoreland Glass emerged in 1889 from what had been Specialty Glass Company. Westmoreland is well known for its novelty glass containers, to be filled with colored candies and shaped like cars, lanterns, trains, rabbits and numerous other fun shapes. They also took out patents for glass violin bottles. More importantly for the Corn Flower story, they were known as one of the largest glass-decorating firms in America through the 1920s and 1930s. At that time they had twenty-five or more cutters on staff. In 1921 George West, one of the company's principals, declared Westmoreland to be "The largest decorating factory in the United States." Many of their blanks from this era were fairly plain lending them to decoration, and perhaps to the notice of W.J. Hughes.

Westmoreland #1503-1 Creamer & #1503-2 Sugar, crystal, 2-1/2", pair $55. *Courtesy of Brenda Beckett.*

West Virginia Glass Specialty (1930-1987)
(Huntington Glass, Louie Glass, Ludwick Glass,
Ludwig Glass, Weston Glass)

It seems appropriate to have the glass companies of Louie Wohinc appear last in this survey of glass suppliers to W.J. Hughes Corn Flower. While apparently not a

major supplier to Hughes in the 1930s, huge amounts of glass were to be purchased from the Wohinc glass concerns for many years beyond 1950.

Austrian-born, Louie Wohinc began working in glassworks at the age of nine. In 1905, while in his late teens, he came to America to work in a glasshouse with an uncle. His story is typical of a glass worker of that era, finding him wandering from factory to factory, staying for some months here, or a couple of years there. What differentiated him from most others was his incredible drive and determination. Graduating to managing plants and working in sales in the 1920s, Wohinc started his own glass shop in 1926, the Louie Glass Company. Over the following years Louie Wohinc expanded his glass empire by forming the West Virginia Glass Specialty Company in 1930, Ludwick Glass in 1940, Weston Glass in 1944, and Huntington Glass in 1945. He was a regular glass-manufacturing baron!

By controlling a clutch of glassworks Wohinc could shift work on moulds back and forth between locations to suit production schedules. We know that the same items were produced at several of his companies. His factories were known for production of a wide variety of beverage wares such as water sets, punch sets, decanters and cocktail shaker sets, as well as console sets and vases. The recognizable Louie style often includes horizontal rings or ribs, handles that are unattached at the top, and jauntily pulled lips on pitchers. Collectors also know well the large variety of crimped-top bud vases and ball vases that were cut in abundance with the Hughes Corn Flower.

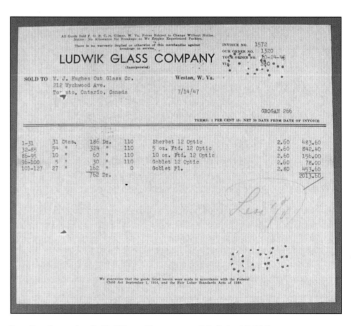

Invoice from Ludwik Glass Company, 14 July 1947 for stemware.

Although no items have been identified in the advertising photographs as West Virginia Specialty Glass, blanks were acquired from them starting as early as 1939.

West Virginia Specialty #1106 Water Jug, crystal, 7-3/4", $75.

Purchases from West Virginia Specialty are verified in some of the few existing business checks from W.J. Hughes. An order for $177.21 worth of merchandise was placed in 1939; an order for $44.55 was place in 1941; and an order for $1368.97 was placed in 1942. Unfortunately no invoices indicating what items were purchased in these early years have been discovered. An invoice does, however, exist from October 1947 for items sold to W.J. Hughes Cut Glass Co. from one Ludwig Glass Company of Weston, West Virginia. This bit of wordplay on the name Louie thinly disguised yet another Louie Glass Company factory. This large order was for an assortment of some 762 dozen stems.

When Louie Wohinc passed away, in 1950, his daughter, Margaret, took over the running of his glass empire. She successfully guided the Wohinc glass works for over two decades, still producing many of the same shapes that were made in the 1930s.

Corn Flower Research

We are beginning to more fully realize the importance of the historical ties between Hughes Corn Flower and most of the major American glass manufacturers of the twentieth century. Much has been revealed about Corn Flower over the past handful of years. Many additional companies who supplied glass and the names of their blanks have been positively identified. A great part of this success has been thanks to the large international community of glass collectors and researchers, so many who freely share their vast knowledge. It is only with such collaboration that we can hope to unravel yet more of the mysteries of this most exciting period of glass making.

Part II

Corn Flower on Color
(1916-1951)

Amber

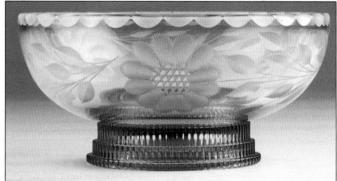

Duncan Miller "Terrace" Bowl, crystal
with amber foot, 4-3/4", $75.

Central Glass Works #2019 Bonbon, 2-handled, 7", $150.
Note that New Martinsville also made the identical bowl with
the same #2019 item designation! The CGW amber is darker.
Courtesy DCMA.

Paden City #890
"Crow's Foot" Console
Bowl, 11-1/2", $275.
Courtesy DCMA.

New Martinsville #42
"Radiance" Footed Bowl,
flared, 10", $250.

Cambridge #411 Comport, flared,
11", $250. *Courtesy DCMA.*

Console Bowl, 10", $250.

New Martinsville #103
Candy Box, 6-1/2",
$300. *Courtesy DCMA.*

Tiffin #14185 Cream and Sugar, 4-1/2",
$155. *Courtesy DCMA.*

Cream and Sugar, 2-3/4", $175. *Courtesy DCMA.*

U.S. Glass Co. #8404
Decanter, 9-1/2", $225.
Courtesy of Brian J. Wing.

New Martinsville #37
"Moondrops" Miniature Cream,
Sugar & Tray, set $225.
Courtesy of Brian J. Wing.

Cambridge #933 Cup and Saucer, 3-1/2", $75.

Duncan Miller #28 Ice Bucket, 6", $350. *Courtesy DCMA.*

Tiffin #14194 Covered Tankard, 2 qt., 11-3/4", $350. *Courtesy DCMA.*

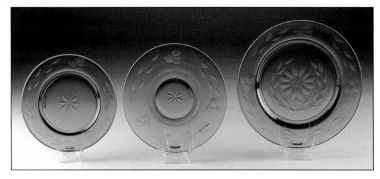

Sherbet and Salad Plates, 5-3/4", $35; 6-1/2",
$35; 7-1/2", $55. *Courtesy DCMA.*

New Martinsville #34 "Addie" Plate, 6",
$35. *Courtesy of John Lovell.*

New Martinsville #34 "Addie" Plate,
8", $50. *Courtesy DCMA.*

Westmoreland #1903 Plate,
scalloped edge, 8-1/2", $75.
Courtesy DCMA.

34

New Martinsville #35 Compote, 5-1/4", $85; #35 "Fancy Squares" Plate, 7-1/2", $55. *Courtesy DCMA.*

Plate, Octagonal Plate, Salad, 7-1/2", $50. *Courtesy DCMA.*

Duncan Miller Relish, 6-part. 11-1/2", $250. *Courtesy DCMA.*

New Martinsville #38 "Hostmaster" Plate, 14-1/2", $250. *Courtesy DCMA.*

Cake Plate, 10-1/2", 4-footed. $175. *Courtesy DCMA.*

New Martinsville #42 "Radiance" Plate, 14", $200. *Courtesy DCMA.*

Tiffin Tray, center-handled, 9-3/4", $225.

Paden City #700 "Simplicity" Server, center-handled oval, 14", $250. *Courtesy DCMA.*

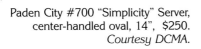

Tiffin #15011 Sundae, 4", $45; Goblet, 8", $75.
Courtesy of Brenda Beckett.

Lancaster Saucer Champagne, 5-1/4",
$55. *Courtesy DCMA.*

New Martinsville #42 "Radiance"
Ball Vase/Punch Bowl, 9", $300.

Duncan Miller, Relish, 3-part
section, 10", $150. A pair of
two of these are pictured in
the 1938 Corn Flower catalog.
Courtesy of Brian J. Wing.

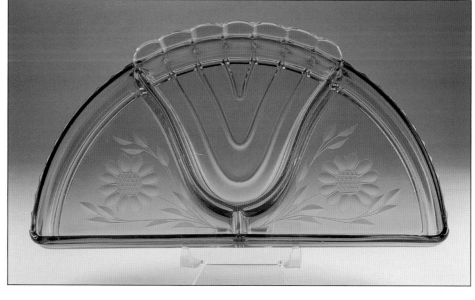

Amethyst

Tiffin #011
Sundae, 4", $75.

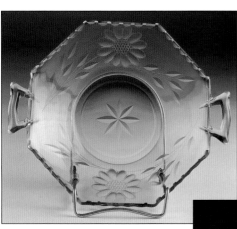

Heisey Hawthorne #1229
"Octagon" Mint, 6", $185.

Hawthorne #1203 "Flat Panel Octagon" Console
Bowl, 12", $400. *Courtesy DCMA.*

Heisey Hawthorne #1229 "Octagon" Jelly, 5-1/2", $185. *Courtesy DCMA.*

Tiffin #14185 Table Tumbler, 4-1/2",
$100. *Courtesy DCMA.*

Blue

Console Bowl, 10", $300. *Courtesy DCMA.*

Tiffin #8076 Sky Blue Berry Bowl, open work, 10", $400. *Courtesy DCMA.*

Tiffin #8307 Sky Blue Bonbon & Cover, 5", $450. *Courtesy DCMA.*

Cream and Sugar, 3-1/4", $300.

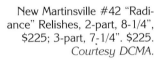

New Martinsville #42 "Radiance" Relishes, 2-part, 8-1/4", $225; 3-part, 7-1/4". $225. *Courtesy DCMA.*

Sherbet Plate, 6-1/4", $55; Salad Plate 7-1/2", $75. *Courtesy DCMA.*

Tiffin #14194 Sky Blue Covered Tankard, 2 qt., 11-1/2", $450. *Courtesy DCMA.*

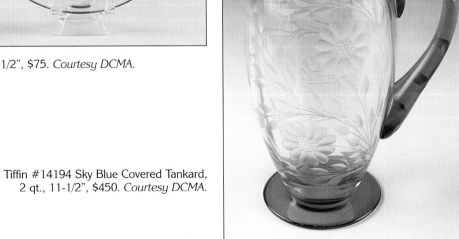

Tiffin #14185 Twilight Table Tumbler, 4-1/2", $100. *Courtesy of Brian J. Wing.*

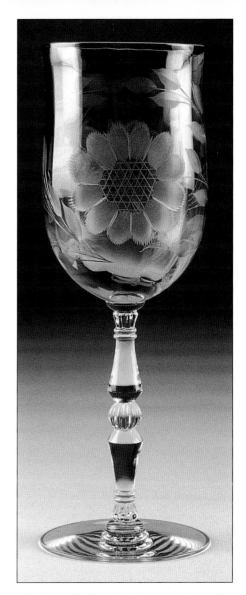

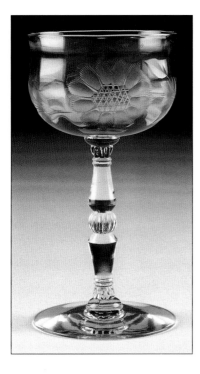

Tiffin #15033 Twilight Cocktail, 5", $125. *Courtesy DCMA.*

Tiffin #15033 Twilight Goblet, 8-1/4", $125. *Courtesy DCMA.*

Tiffin #15001 Sky Blue Goblet, 7" $100; Sundae, 4-1/4, $75. *Courtesy DCMA.*

Juice Tumbler, 4" footed, $85; Water Tumbler 5" footed. $125. *Courtesy DCMA.*

Gold Trim

Tiffin #354 Table Tumbler, 4", $30; #112
Squat Jug, 7-1/4", $125.

Tiffin #336-2 Cake Plate, handled, 11", $150.
Courtesy DCMA.

Tiffin #020 Goblet, 6-1/4", $35; Sundae, 3-1/2", $25.
Courtesy of Brian J. Wing.

Tiffin #020 Oyster Cocktail, 3-3/4", $30.
Courtesy DCMA.

Green

Tiffin #15151 Art Basket, 6-1/4", $225. *Courtesy DCMA.*

Console Bowl, 10", $250. *Courtesy of Brenda Beckett.*

Duncan Miller #6 Flower Bowl, flared 12", $300. *Courtesy of Brenda Beckett.*

Duncan Miller #6 Flower Bowl, cupped, 8-1/2", $300. *Courtesy DCMA.*

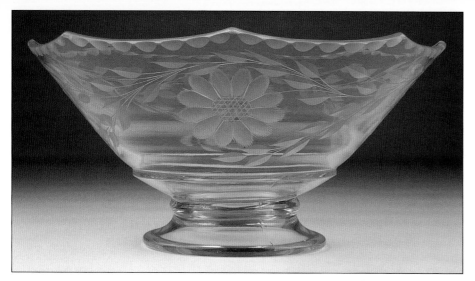

Tiffin #345 Footed Comport, low, 9-1/2", $250. *Courtesy of Brenda Beckett.*

Bowl, 3-footed, scroll feet, 5", $225. *Courtesy of Brenda Beckett.*

Duncan Miller #3 "Three Leaf" Candlesticks, 2-3/4". Note bud variant instead of Corn Flower cut, $225. *Courtesy of John Lovell.*

Tiffin #101 Candlesticks, 1-lite, 3". With bud cut, $200. *Courtesy DCMA.*

44

Lancaster #355
Candlesticks, 1-lite, 3".
With bud cut, $200.
Courtesy DCMA.

New Martinsville #103 Candy Box, 6-1/2",
$300. *Courtesy DCMA.*

Tiffin #330 Bonbon & Cover, high-footed, 8-1/2",
$275. *Courtesy DCMA.*

Tiffin #9557 Candy
Jar, 1/2 lb, crystal with
green trim, 8-1/4",
$275. *Courtesy
DCMA.*

Boat, Bonbon, 7-3/4", $175.
Courtesy of Brenda Beckett.

Tiffin #348 Compote, 7-1/2", $250.
Courtesy of Dave and Hilda Proctor.

Tiffin #315 Compote, high foot, 7-1/2", $250. *Courtesy DCMA.*

Cream and Sugar, 3-1/4", $225. *Courtesy of John Lovell.*

Plate, octagonal salad, 7-1/2", $65. *Courtesy DCMA.*

New Martinsville #38 "Hostmaster" Plate, 14", $275.

Plate, 6" Sherbet, $35; 7-1/2" Salad, $60. *Courtesy DCMA.*

Tiffin #330 Center-handled Cake Plate, $225.

Westmoreland #1902 Footed Salver, low, 10",
$225. *Courtesy DCMA.*

Tiffin #320 Reflex Green Whipped Cream, 6-1/2".
Bud cut on liner, $250. *Courtesy DCMA.*

Mayonnaise with Underplate and Ladle, 6-1/4". Bud cut
on mayo and underplate, $250. *Courtesy DCMA.*

Heisey #1226 "Octagon" Mayonnaise, footed, 5-1/2", $225.

Pitcher, crystal with green trim, $300.

New Martinsville #36 Relish, 3-compartment, 8", $225. *Courtesy of Brenda Beckett.*

Tiffin #8896 Relish, 3-part , 9-3/4", $300. *Courtesy of Brenda Beckett.*

49

Goblets, crystal with green trim, 4-1/2",
$65; 5-1/4", $75. *Courtesy DCMA.*

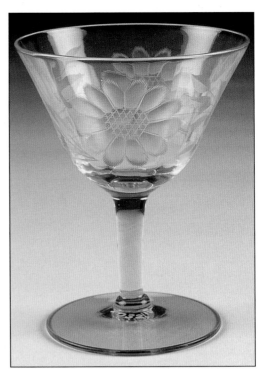

Lancaster Saucer
Champagne, 4-1/2"
$55. *Courtesy DCMA.*

Tiffin #15001
Sundae, crystal with
green trim, 4-1/4",
$55. *Courtesy
DCMA.*

Tiffin #15040 Goblet, crystal
with green trim, 7", $75.
Courtesy DCMA.

Tiffin #15011 Goblet,
crystal with green trim, 6",
$85. *Courtesy DCMA.*

Tiffin #15040 Sundae,
crystal with green trim,
3-3/4", $50. *Courtesy
of Brenda Beckett.*

Tiffin #001 Sundae, Rose with green trim and festoon optic, 4-1/4", $100. *Courtesy DCMA.*

New Martinsville Tumbler, 3-1/2", $45; #37 "Moondrops" Mug, 6-3/4", $225. *Courtesy DCMA.*

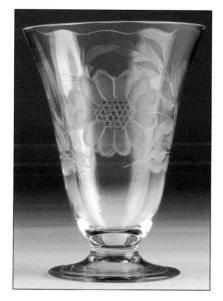

Tiffin Table Tumbler, 4-3/4", $50.

Tiffin #020 Table Tumbler, 4-1/2", $45; Oyster Cocktail, 2-3/4", footed, $65. *Courtesy DCMA.*

Tiffin #6471 Vase, 9". With bud cut around collar, $300. *Courtesy of John Lovell.*

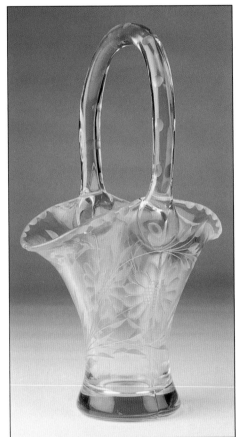

Lancaster R1830-1 Bowl, 3-footed, 6", $175

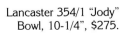
U.S. Glass Co. #9574 Basket, 11",
$350. Though marketed as Tiffin, it
was actually made at another U.S.
Glass factory, either "G" or "K" in
Pittsburgh. *Courtesy of Brenda
Beckett.*

Lancaster 354/4 "Jody" Bowl, 10", $275. *Courtesy DCMA.*

Lancaster 354/1 "Jody"
Bowl, 10-1/4", $275.

Duncan Miller #6 Flower Bowl, 12", $300. *Courtesy of Brenda Beckett.*

Lancaster T1831 Rose Bowls, 9", $225; 6", $200. *Courtesy DCMA.*

Lancaster T1894-1 Bowls, 2-handled, 7-1/2",$150; 11",$225; 7-1/2", $150. *Courtesy DCMA.*

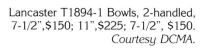

Heisey #1229 "Octagon" Mint, 6", $150. *Courtesy DCMA.*

Bowls, 3-footed, scroll feet, 7", $225; 4", $150. *Courtesy DCMA.*

Duncan Miller #31 Nappy, 3-toed, 8", $250.

Console Bowl, 12", $250.

Tiffin #1502 Finger Bowl, 4-1/2", $150. *Courtesy DCMA.*

Heisey #1252 Flamingo "Twist"
Mint, 3-cornered, 5-1/2", $150.
Courtesy DCMA.

Heisey #355 "Quator" Bonbon,
4", $150. *Courtesy DCMA.*

Tiffin #345 Rose Pink Comport,
low-footed, 9-1/2", $250.

Tiffin #330 Rose Pink Comport, low-
footed, 8-3/4", $250. *Courtesy DCMA.*

Rose Pink Comport, 8",
$250. *Courtesy DCMA.*

Rose Pink Comport, 8-1/2",
$250. *Courtesy DCMA.*

Heisey #112 "Mercury"
Candlestick, 3". Patent
No. 70558, pair $225.

Beaumont Lemon Tray, center-handled, 6",
$150. *Courtesy DCMA.*

Lancaster #98 Nappy and
Cover, three-toed, 6", $275.

Tiffin #330 Low-footed Bonbon and
Cover, 6", $300.

Tiffin #330 Conic Candy Jar
and Cover, 8", $250.
Courtesy of Brenda Beckett.

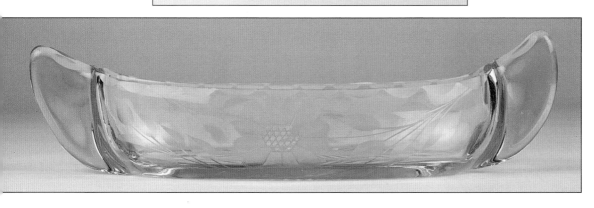

Boat Bonbon,
8-1/2", $200.
*Courtesy of Ken and
Terri Farmer.*

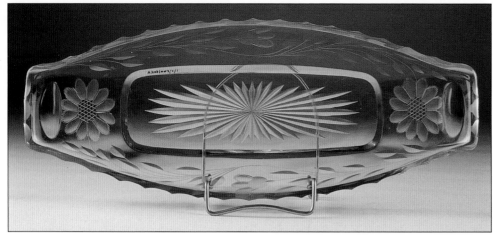

Heisey #1185 Flamingo "Yeoman" Celery, 12". *Courtesy DCMA.*

Tiffin #151 Celery, 12", $225. *Courtesy of Brenda Beckett.*

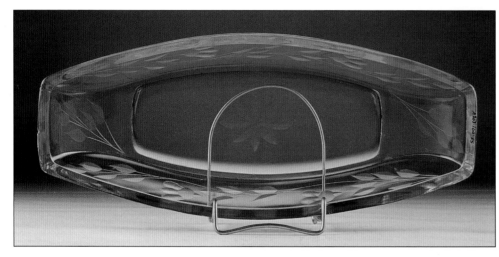

Celery, 11-1/4", $225. *Courtesy DCMA.*

Tiffin #315 Comport, low-footed, 5", $225.

Tiffin #319 Comport, high-footed, 7", $250.
Courtesy DCMA.

Tiffin #14185 Cream & Sugar, 4-1/4", pair $175. *Courtesy DCMA.*

Lancaster #737 Cream & Sugar, 3-3/4",
pair $200. *Courtesy DCMA.*

Lancaster #879 Creamer, 3-1/4",
$100. *Courtesy DCMA.*

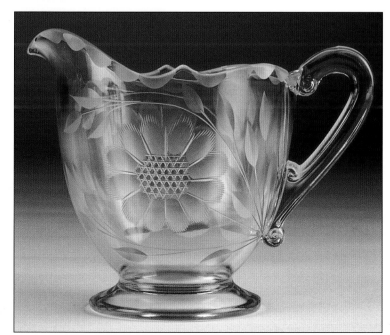

59

U.S. Glass Co. #8404 Decanter. 9-1/2", $225; Wine, 2-3/4, $50. *Courtesy of Brenda Beckett.*

Lemon Dish & Cover, 5", $250. *Courtesy of Brian J. Wing.*

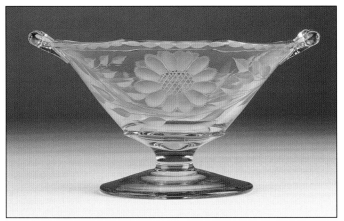

Heisey #1226 "Octagon" Mayonnaise, footed, 5-1/2", $225.

Heisey #1401 Flamingo "Empress" Pickle & Olive, 2 compartments, 13", $200. *Courtesy DCMA.*

Mayonnaise with Liner & Ladle. With bud cut, 6",
$200. *Courtesy DCMA.*

Plates, 6", $30; 6-1/2", $30; 7-1/2", $55. *Courtesy DCMA.*

Tiffin #14185 Mayonnaise with Liner, 5", $200. *Courtesy DCMA.*

Plate, Salad, octagonal, 7", $60. *Courtesy DCMA.*

Westmoreland Tray, center-handled, 10", $175. *Courtesy DCMA.*

Tiffin #330 Tray, center-handled, 10", $225. *Courtesy DCMA.*

Lancaster T885 Sandwich Tray, 11", $225. *Courtesy DCMA.*

Tiffin #336-2 Cake Plate, handled, 11-1/4", $175.
Courtesy DCMA.

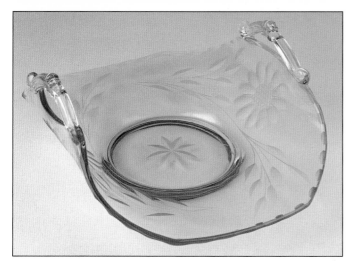

Lancaster Bonbon, 2-handled, 7", $175.
Courtesy of Brenda Beckett.

Tiffin #112 Rose Pink Jug, 7-1/4", $350.
Courtesy DCMA.

New Martinsville #38 Relish, 3-compartment,
9", $175. *Courtesy DCMA.*

Louie Cordial, 3-1/2", $75; Water
Goblet, 6-1/2", $65. *Courtesy DCMA.*

Heisey #1184 "Yeoman" French
Dressing Boat & Plate, 7", $325.

Louie #625 Cocktail, 5",$75; #625 Wine, 5-1/2", $75. *Courtesy of John Lovell.*

Tiffin #15011 Goblet, 8-1/4", $75; Wine, 6", $85; Cordial, 5", $95. *Courtesy DCMA.*

Tumbler, 9-sided, 3-1/4", $55. *Courtesy DCMA.*

Louie Goblet, 5", $55; Sherbet, 3-1/2", $35.

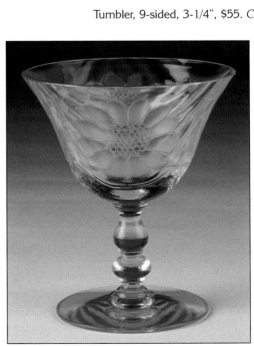

Tiffin #001 Sundae, Rose with green trim and festoon optic, 4-1/4", $100.

Tiffin #580 Tumbler, 3-3/4", $50.
Courtesy DCMA.

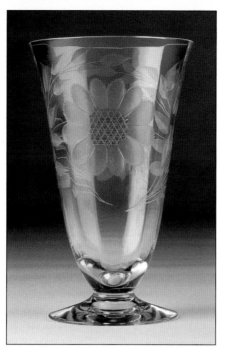

Tiffin #020 Ice Tea, footed, 5-3/4", $65.

Tiffin #020 Table Tumbler, Rose with green trim and festoon optic, footed, 4-3/4, $110. *Courtesy DCMA.*

Tiffin #14185 Rose Pink Table Tumbler, 4-1/2", $55. *Courtesy DCMA.*

Tiffin Tumbler, Juice, 3-3/4", $40. *Courtesy DCMA.*

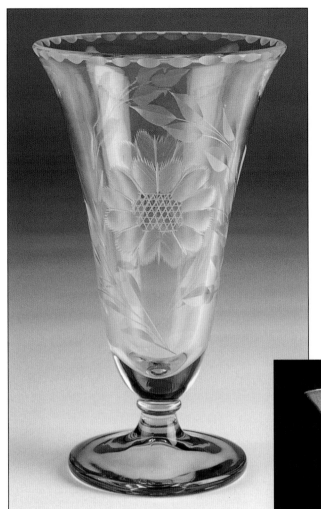

Vase, 7", $225.

Tiffin #2 Rose Pink Vase, 6", $200.
Courtesy DCMA.

Tiffin #14185 Bud Vase, 10",
$155. *Courtesy DCMA.*

66

Red

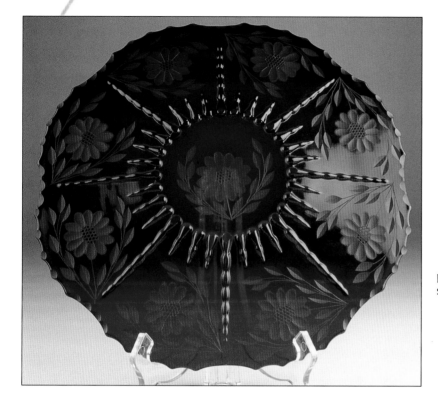

New Martinsville #42 "Radiance" Plate, 11-1/2", $450. *Courtesy of John Lovell.*

Indiana #6828 Candlestick, 2-lite, fired-on red, pair $225.

Polish Pilsener, 8-1/2", $65.
Courtesy of Mike Lutes.

Mid-Atlantic
Tumblers: Juice,
3-1/4", $35; Water,
4", $40; Ice Tea,
4-1/2", $50.
Courtesy DCMA.

Vaseline

Tiffin #15151 Canary Art Basket, 6-1/4",
$225. *Courtesy DCMA.*

Tumbler, 4-1/2", $55.
*Courtesy of Brian J.
Wing.*

Plate, 7-1/2", $65. *Courtesy DCMA.*

Tiffin #8076 Berry Bowl, open work, 11", $350.
Courtesy DCMA.

Yellow

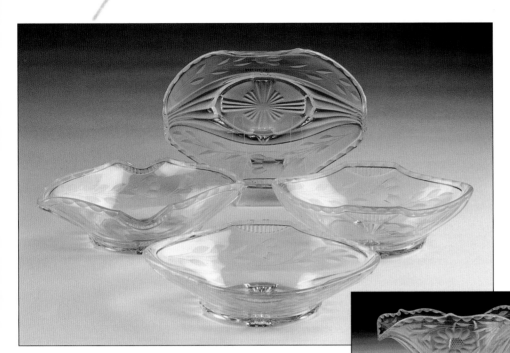

Lancaster #353 "Jody" Bowls, 6-1/2"
to 7", $125 each. *Courtesy DCMA.*

Lancaster #767/1 Bowl, crimped, 3-footed, 6-1/2", $125;
#767/3 Bowl, 3-footed, 6", $125. *Courtesy DCMA.*

Lancaster #354/4 Bowl, 11"; #354/1 Bowl, 11",
$250 each. *Courtesy DCMA.*

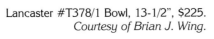

Lancaster #T378/1 Bowl, 13-1/2", $225.
Courtesy of Brian J. Wing.

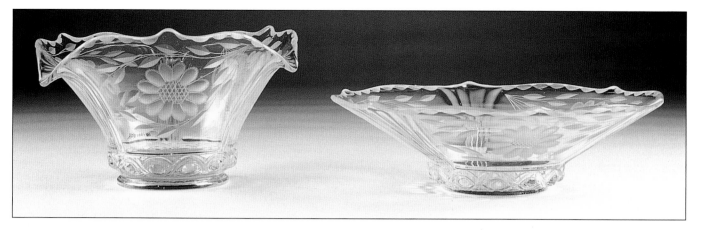

Lancaster #T1377 Bowl, crimped, 6". #T378 Bowl, 8", $150 each. *Courtesy DCMA.*

Lancaster #T1894-1 Bowl, 2-handled, 5-3/4", $125. *Courtesy DCMA.*

Lancaster #R1786/4 Plate, 2-handled, 12-1/2",
$225; #R1786/1 Bowl, 2-handled, 11", $225.

Heisey #1229 "Octagon" Sahara Mint, 6", $155.
Courtesy DCMA.

Tiffin #342 Candlesticks, 4", pair
$225. *Courtesy of Brenda Beckett.*

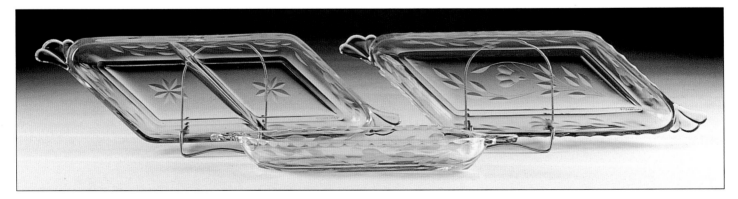

Lancaster #833 Candlesticks, 3",
pair $175. *Courtesy of Brenda
Beckett.*

Heisey #1401 "Empress" Sahara Pickle & Olive, 2 compartments, 13",
$175; Celery Tray, 10", $150; Celery Tray, 13", $175. *Courtesy DCMA.*

Lancaster #879
Cream & Sugar,
3-1/4", Pair $200.

Tiffin #348 Compote, low footed 6", $200.

Lancaster #T1831 Tray, 4-footed, 8-1/4", $125. *Courtesy DCMA.*

Lancaster #T1831 Tray,
4-footed, 7-1/2", $125;
#T1831/7 Tray, 4-
footed, 11", $155.
Courtesy DCMA.

73

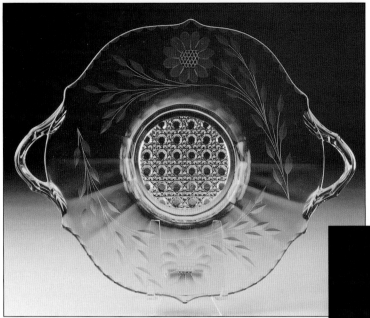

Lancaster #T899/4 Topaz Tray, 2-handled, 11", $175. *Courtesy DCMA.*

Lancaster Topaz Tray, 2-handled, 12-1/2", $175. *Courtesy DCMA.*

Tiffin #14185 Sundae, 4-1/2" $50; Saucer Champagne, 6-1/4", $85. *Courtesy of Brenda Beckett.*

Lancaster Topaz #T885 Sandwich Tray, center-handled, 11", $225. *Courtesy DCMA.*

74

Plates: 6", $35; 7-1/2", $50. *Courtesy DCMA.*

Table Tumbler, footed, 4-3/4", $55; Oyster Cocktail, footed, 3-3/4", $45. *Courtesy of Brenda Beckett.*

Tiffin #196 Goblet, 6-1/2", $65; Saucer Champagne, 4-1/2", $60; Oyster Cocktail, 3-3/4", $50; Sundae, 3-1/2", $40. *Courtesy DCMA.*

Tiffin #5831 Mandarin Footed Bowl, 4-3/4", $125. *Courtesy DCMA.*

Part III

Corn Flower on Early Crystal
(1916–1951)

Crystal

Tiffin #0142 Nappy, 1-handled, 5", $45. *Courtesy DCMA.*

Heisey #1483 "Stanhope"
Mayonnaise, 2-handled, $50.
Courtesy DCMA.

Heisey #1229 "Octagon" Mint, 6",
$40. *Courtesy DCMA.*

Bowls, 3-toed: 5", $45; 6", $55; 7", $65. *Courtesy DCMA.*

Heisey #1183 "Revere" Mint,
2-handled, $30.

Heisey #1184 "Yeoman" Preserve,
5-1/2", $40; Oval Fruit, 9-1/2",
$65. Patent number 50666.
Courtesy DCMA.

77

Cambridge Finger
Bowl, 4-1/4", $40.

Paden City #330 "Cavendish" Handled
Nappy, 7-3/4", $35. *Courtesy DCMA.*

Cambridge #325 Bonbon, 2-handled,
7-1/2", $35. *Courtesy of John Lovell.*

Duncan Miller #31 Nappy, 3-toed, 8",
$75; 6", $65. *Courtesy DCMA.*

Bowls, 4-1/2", $40; 8", $65.
Courtesy DCMA.

Console Bowl, 11-1/2", $95.
Courtesy of Brenda Beckett.

Paden City #221 "Maya" Bowl, 2-handled, 9-1/2",
$65. *Courtesy of Brian J. Wing.*

Bowl, 6", $40. Note the unusual dot decoration
just below the rim. This is also seen on tankard
pitchers c.1948. *Courtesy DCMA.*

79

Bowls, Fruit Intaglio bases, 4-3/4",
$20; 8", $35. *Courtesy DCMA.*

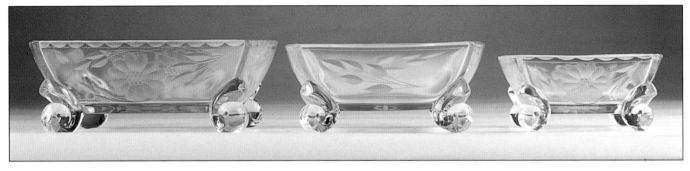

Duncan Miller #152-86 Bon-Bon, 5-1/2", $40; #152-85 Mint or Jelly, 4-1/2",
$35; #152-84 "Patio" Nut, 4-footed, 3-1/2", $30. *Courtesy of Brian J. Wing.*

Duncan Miller #152-106 "Patio" Rose Bowl, 4-footed, 5". $40.
Courtesy of Brenda Beckett.

New Martinsville #42 "Radiance" Footed Bowl, flared, 10". With
original label, $75. *Courtesy of Brenda Beckett.*

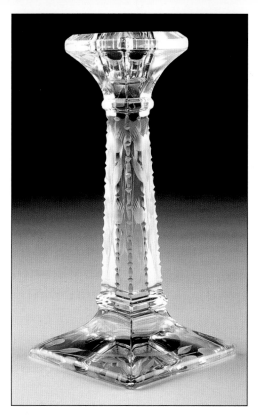

Duncan Miller #114-121 "Canterbury" low Candlestick, 3-1/4", pair $55.

Heisey #21 "Aristocrat" Candlestick, 1-lite, 7", pair $75. Patent #41590. *Courtesy DCMA.*

Paden City #777 "Comet" Candlesticks, 1-lite, 4-3/4", pair $75.

Paden City #330 "Cavendish" Candlestick, 6-1/2", pair $85. Each has a hole for a prism. *Courtesy of Brenda Beckett.*

Candlesticks, 2", pair $55. *Courtesy DCMA.*

New Martinsville #4453
Candlestick, 1-lite, 6", pair $75.

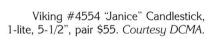

Viking #4554 "Janice" Candlestick,
1-lite, 5-1/2", pair $55. *Courtesy DCMA.*

Tiffin #15360 Candelabrum,
2-lite, 6-1/2", pair $75.

Heisey #103 "Cupped Saucer"
Candlesticks, 1-lite, 2-1/2", pair
$55. *Courtesy of Lynda
Edwards.*

Tiffin #5831 Candelabrum, 2-lite, 7", pair $65.

Viking #4536 "Janice" Candlesticks, 2-lite, pair $55. *Courtesy DCMA.*

Indiana Candlesticks, 2-lite, 5-1/2", pair $85.

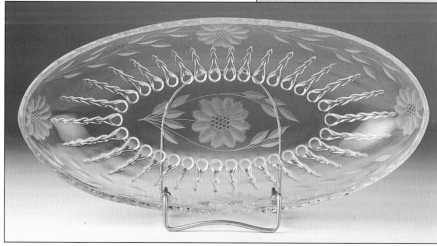

New Martinsville #42 "Radiance" Celery, 10-1/4", $45. *Courtesy of Brenda Beckett.*

Paden City #412 "Crow's Foot" Candy Dish & Lid, 2-part, $75. *Courtesy of Brian J. Wing.*

Heisey Candy & Lid, 6", $95. *Courtesy DCMA.*

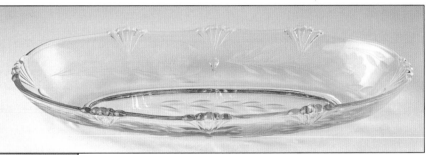

Paden City #890 "Crow's Foot" Celery, 11-1/2", $45. *Courtesy of Brenda Beckett.*

Paden City #890 "Crow's Foot" Candy Dish & Lid, 3-part, $85. *Courtesy of Brenda Beckett.*

Heisey Celery, 10-1/4", $55. *Courtesy DCMA.*

Central-Imperial #148 Celery, 10-1/2", $45.

Tiffin #15181 Celery, 8-1/4", $55.

Central-Imperial #148 Celery, 11-1/4", $55.
Courtesy of Lynda Edwards.

Heisey #4044 "New Century" Celery Tray, 13", $75. *Courtesy DCMA.*

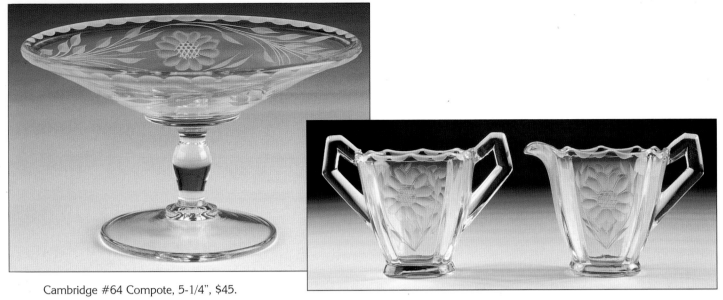

Cambridge #64 Compote, 5-1/4", $45.

Paden City #411 "Mrs. B" Cream & Sugar, 3-1/8",
pair $75. *Courtesy DCMA.*

Tiffin #082 Grape Fruit, pair, 6-1/4", $55
each. *Courtesy of Brenda Beckett.*

Tall Compote, 5", $45. *Courtesy
of Brenda Beckett.*

Imperial #588 "Pie Crust" Cream, Sugar, individual,
2-3/4" & Tray, 7-1/4". Manufactured by Imperial from
the mid-1920s through to 1945, set $65.

Paden City #221
"Maya" Cream & Sugar,
3-1/4", pair $65.

Paden City #555 "Gazebo"
Cream & Sugar, 3", pair $65.

Tiffin #036 Cream &
Sugar, 3", pair $95.

Indiana Cream & Sugar, 4-1/4",
pair $65. *Courtesy of Brenda
Beckett.*

Cream & Sugar, 3-1/2", pair $65. *Courtesy DCMA.*

Westmoreland #1900-12 Cream & #1900-16 Sugar, 2-1/2", pair $65. *Courtesy of Brenda Beckett.*

Imperial Tumbler, 4", $20; #451 "Georgian" Jug, bulbous, 7", $85. *Courtesy DCMA.*

Tiffin Tumbler, 4-1/2", $20; #14194, Covered Tankard, 1 qt., 11", $175. *Courtesy of Brenda Beckett.*

Tiffin #0275 Ice Tea Jug & Cover, 10-1/2",
$175. *Courtesy DCMA.*

Jug, bulbous, 7-1/2", $75.

Tiffin #017 Jug, 7", $125.
Courtesy DCMA.

Tiffin #02, Table Tumbler, 3-3/4", $25; #011 Jug, 2-qt., 8-1/2", $125. *Courtesy of Brenda Beckett.*

West Virginia Specialty Pitcher, 9", $65. *Courtesy DCMA.*

Heisey Jug, 7", $85. *Courtesy of John Lovell.*

Heisey Pitcher, 5", $55. *Courtesy DCMA.*

Plate, 6". Note early gang cut, grooved edge decoration. Thought to date from the teens or 20s, $30.

Plate, 8-1/4", with bud decoration, $40.

Cheese Plate with indent, 9-3/4". Missing the compote. Note overcut star central portions and extra details in decoration encircling the center, $55. *Courtesy DCMA.*

Paden City "Nerva" Plate, 11-3/4", $65.

Paden City #555 "Gazebo" Plate, 11", $65.
Courtesy of Brenda Beckett.

Paden City #220 "Largo" Plate, 15", $75. *Courtesy DCMA.*

Duncan Miller "Terrace" Plate, 15", $75.

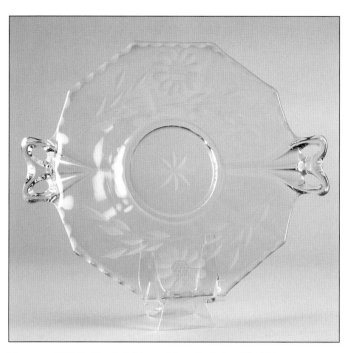

Paden City #555 "Gazebo" Tray, 2-handled, 12-1/2", $55.

Fostoria #2375 "Fairfax" Plate, 2-handled, 8-1/2", $30.

Paden City #215 "Glades" Plate, 2-tab handles, 8", $35. *Courtesy of Brian J. Wing.*

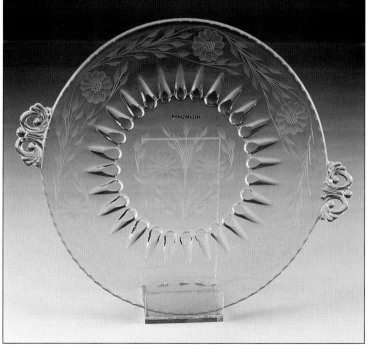

Viking #4462 Plate, 2-handled, 13", $50. *Courtesy DCMA.*

Central-Imperial #1480/C261 Cake Plate, 2-handled, 11", $50. *Courtesy DCMA.*

Plate, 2-handled, 11-1/2", $50.

Beaumont Tray, center-handled, 8-1/2", $65.

Paden City #221 "Spire" Tray, center-handled, 10-3/4", $75. *Courtesy of Brenda Beckett.*

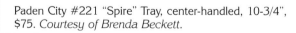

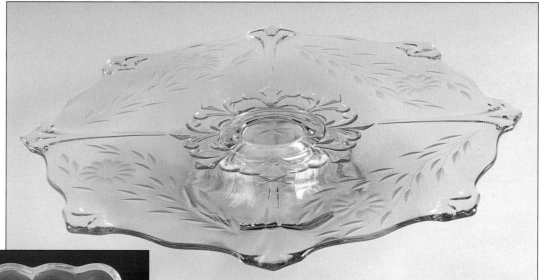

Indiana #607 Pedestal Cake Plate, 13-3/4", $75.

Relish, 2-part, 6-1/2", $25. *Courtesy DCMA.*

Duncan Miller, #115 "Canterbury" Relish, 2-compartment, round, 8", $25.

Paden City Relish, 2-part, square, 7", $30. *Courtesy DCMA.*

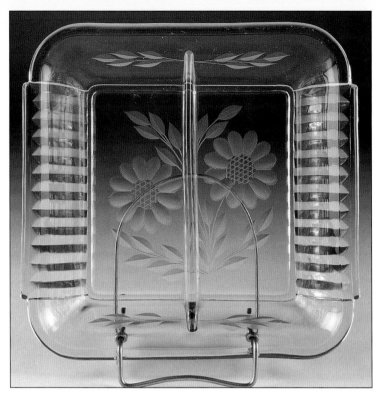

Paden City #215 "Glades" Relish, 2-part square, 7", $30. *Courtesy DCMA.*

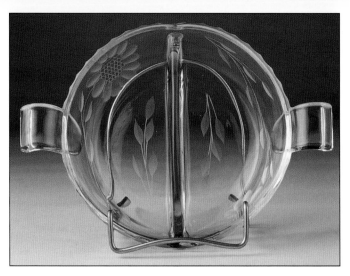

New Martinsville #507 Relish, 2-part, 2-handled, 5", $30. *Courtesy of Brenda Beckett.*

New Martinsville #38 Relish, 3-compartment, 9", $45. *Courtesy DCMA.*

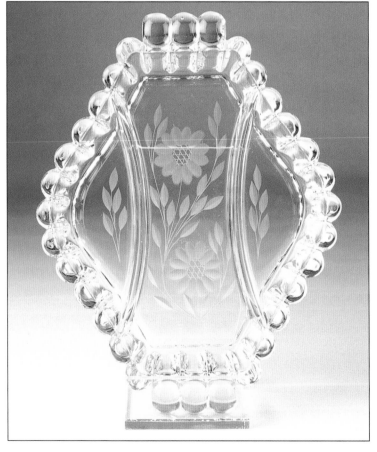

Heisey #1503 "Crystolite" Relish, 3-part oblong, 9-1/2", $35. *Courtesy of Brenda Beckett.*

Heisey #1483 "Stanhope" Triplex Buffet Relish, 2-handled, 14", $65. *Courtesy DCMA.*

Paden City Nerva, Relish, 3-part, 10-3-/4", $50. *Courtesy DCMA.*

Heisey #1483 "Stanhope" Jelly, 3-compartment, 1-handled, 6", $40. *Courtesy of Brenda Beckett.*

New Martinsville #37/4 Relish, 3-part, 2-handled, 12-3/4", $55. *Courtesy DCMA.*

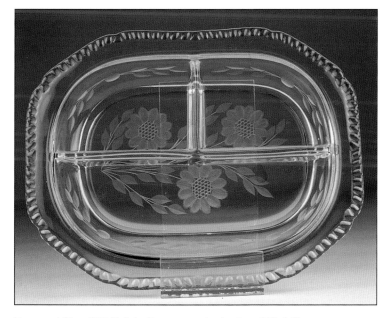

Duncan Miller #52 Relish, 3-part, notched-edge, 10", $45. *Courtesy of Brenda Beckett.*

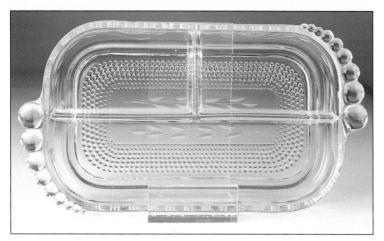

Duncan Miller, #301 "Tear Drop" Relish, 3-compartment, oblong,
12", two styles, $45 each. *Courtesy of Brenda Beckett.*

Relish, 3-part, 9-3/4", $45. *Courtesy DCMA.*

Fostoria #2470 Relish, 3-part,
round-handled, 9-3/4", $45.
Courtesy of Brenda Beckett.

Paden City #777
"Comet" Relish, 3-
part, 7-1/2", $35.

Paden City #215 "Glades" Relish, 4-part, 11", $55.

Duncan Miller Relish, 4-part, 10-3/4", $45.
Courtesy DCMA.

Paden City Relish, 4-part, 13-1/2", $55. *Courtesy of Brenda Beckett.*

Central-Imperial #1480 Relish, 4-part,
2-handled, 11", $50. *Courtesy DCMA.*

Duncan Miller #52 Relish,
notched-edge, 6-part, 10",
$55. *Courtesy DCMA.*

Duncan Miller #52 Relish, notched-edge, 5-part,
10", $45. *Courtesy of Brenda Beckett.*

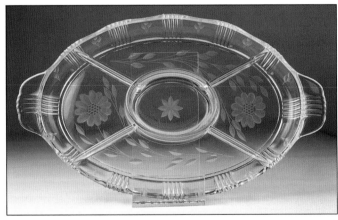

Relish or Buffet Set, 5-part, without lid, 13",
$65. *Courtesy of Brenda Beckett.*

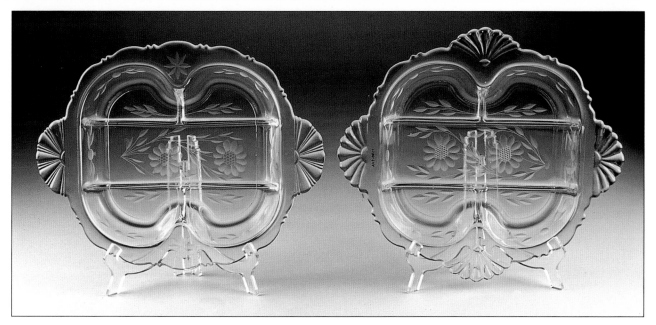

Tiffin #8897 Relish, 5-part, 12-1/4", each $55. *Courtesy DCMA.*

Paden City #220 "Largo" Relish, 5-part, 14", $65. *Courtesy DCMA.*

Tiffin #020 Goblet, 6-1/4", $25; Saucer Champagne, 4-1/2", $20; Cocktail, 4-1/4", $18; Sundae, 3-1/2". $15.

Heisey Saucer Champagne, $25; #1184 Coaster Plate, 4-1/2". Two stems are photographed at a quarter turn difference showing both Corn Flower and bud on the same stem, $25. *Courtesy DCMA.*

101

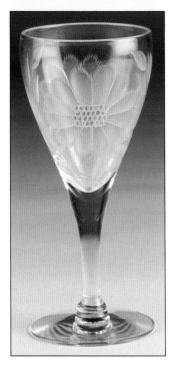

Cordial, 4-1/2", $25.

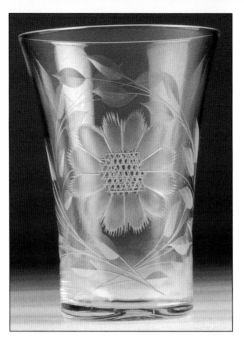

Tiffin #04 Bell Tumbler, 4-1/4", $25.

Cambridge #281
Vase, 6-1/2". $65.

Tiffin #010 Handled
Ice Tea, 5", $30.
Courtesy DCMA.

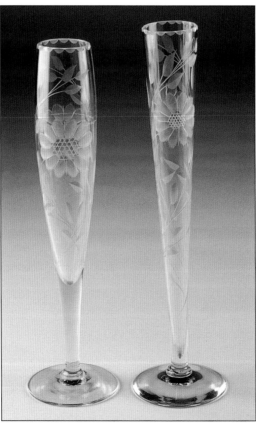

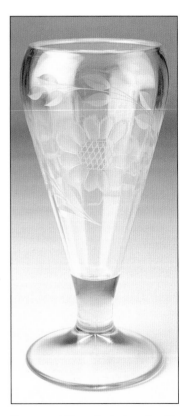

Tiffin #02 Table Tumbler,
3-3/4", $20.

Cambridge #272 Bud Vase, 10", $45; Cambridge
#274 Bud Vase, 10-1/2", $45. *Courtesy DCMA.*

Beaumont Vase, footed, optic, 8",
$55. *Courtesy of Brenda
Beckett.*

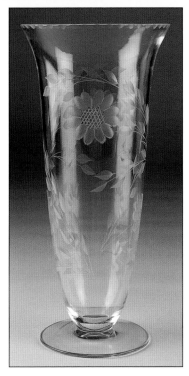

Cambridge #278
Vase, footed, 10-1/2",
$65. *Courtesy DCMA.*

Cambridge #2357 Vases, pair, 8". Corn Flower and bud
cut on each vase. Beading, piecrust edge is an earlier
variant. $75 each. *Courtesy of Brenda Beckett.*

West Virginia Specialty Vase, flip,
12", $75. *Courtesy of Brenda
Beckett.*

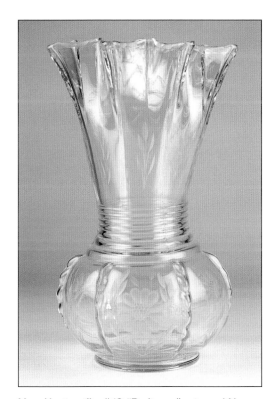

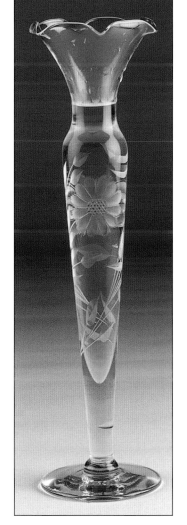

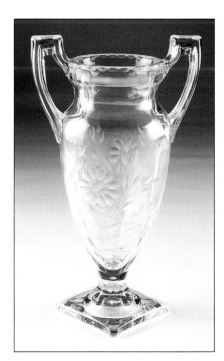

New Martinsville #42 "Radiance" crimped Vase,
10", $95. *Courtesy of Lynda Edwards.*

Cambridge #276 Bud Vase,
10", $45.

Tiffin #15144-1/2 Handled Vase,
10", $200. *Courtesy DCMA.*

Miscellaneous Items

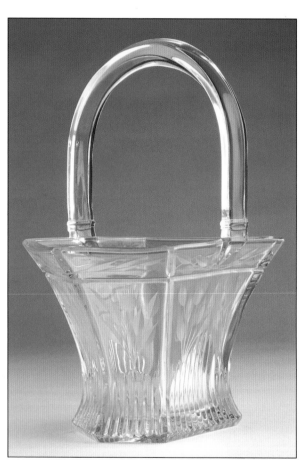

Heisey #461 "Banded Picket" Basket, 9-1/2", $225. *Courtesy of Brian J. Wing.*

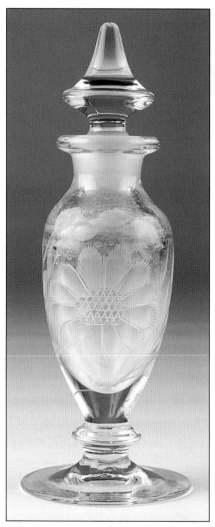

Heisey #491 "Karen" Cologne with stopper, $100. *Courtesy of John Lovell.*

Cocktail Shaker, 9", $75. *Courtesy DCMA.*

Boat Bonbon, 8-3/4", $45. *Courtesy DCMA.*

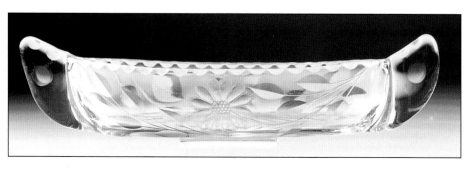

Crystal Paden City #888 "MK" Covered Cheese & Cracker, 12". $85. *Courtesy of Brenda Beckett.*

Tiffin Custard Cup, 2", $15. *Courtesy DCMA.*

Indiana #9 Butter Ball, handled, 6", $45. *Courtesy DCMA.*

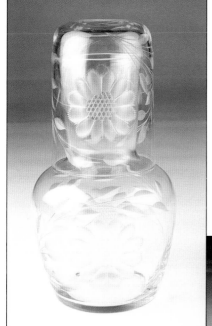

Tiffin #6712 Tumble Up, "Night Cap Set", 6-1/4", $85. *Courtesy of Brenda Beckett.*

Paden City #215 "Glades" Gravy Bowl, 2-spout, 7", $55; Ladle with star cut base & leaves on edge, $25.

Paden City #191 "Party Line" Cordial, 3-1/2", $20; Ice Tub, 6", $125.

Duncan Miller #28 Ice Bucket, 6", $350.
Courtesy of Patrick Doherty.

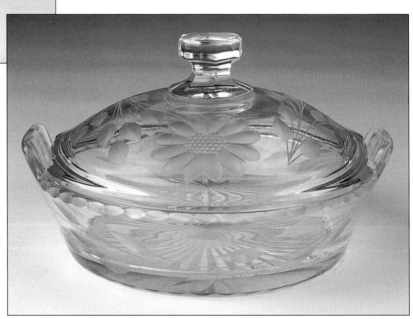

Heisey #1184 "Yeoman" Lemon Dish, 5", bud & flower cuts, $55. *Courtesy DCMA.*

Heisey #1184 "Yeoman" Lemon Dish & Cover, 5", $85.
Courtesy of Brenda Beckett.

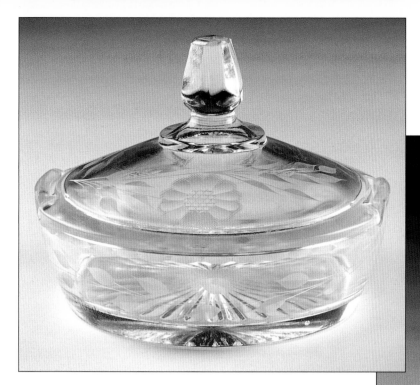

Lemon Dish and Cover, 5",
$85. *Courtesy DCMA.*

Tiffin #179 Oil Bottle, 4-1/2" without
stopper, bud cut, $65. *Courtesy DCMA.*

Marmalade Jar, Lid &
Spoon, 3-3/4", $85.

Tiffin #065 Marmalade Jar, Lid & Spoon,
2-3/4", $85. *Courtesy DCMA.*

Mirror, octagonal, 12", $75.
Courtesy of Brenda Beckett.

Salt & Peppers, 2-1/2", $45. *Courtesy DCMA.*

Heisey #1184 "Yeoman" French Dressing
Boat & Ladle, $85. *Courtesy DCMA.*

Duncan Miller #30 "Pall Mall" Swan, 7", $40.

Paden City Syrup & Metal Lid, 4-1/4",
$85. *Courtesy of Brian J. Wing.*

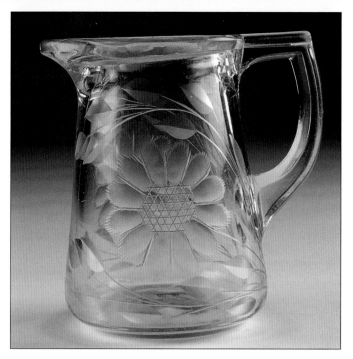

Paden City #198 Syrup, 4-1/4", missing metal lid, complete $85. *Courtesy DCMA.*

Paden City Syrup with Lid, 5-1/4", $65. *Courtesy DCMA.*

Imperial #100 Toothpick Holders, boot-shape, $35 each. *Courtesy DCMA.*

Tiffin #15365 Hostess Tray, large, 14-3/4", $95. *Courtesy of Brenda Beckett.*

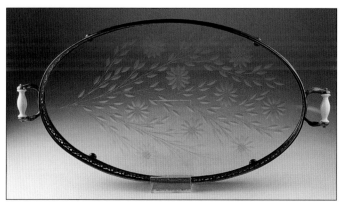

Tray, large serving, 20", $250. This is a Hughes family piece given as a gift by W.J. Hughes, c.1935. Note a variant of the petals with no fringing. *Courtesy of John Lovell.*

Crystal Pattern Groupings

Duncan Miller #301 "Tear Drop" Plate, 2-handled, 13", $45; Cream & Sugar, pair $55. *Courtesy of Brenda Beckett.*

Heisey #1183 "Revere" Jelly, 2-handled, 5", $25; Mint, 2-handled, 6", $25; Bonbon, 2-handled, 6-1/2", $30.

Fostoria #2375 "Fairfax" Bowl, 2-handled 7-1/2", $35; Plate, 2-handled, 8-1/2", $35. *Courtesy of Brenda Beckett.*

Imperial #280/92D "Corinthian" Plate, 14", $55; #280 Bowl, Cupped, 9", $65. *Courtesy of Brenda Beckett.*

Imperial "Crocheted Crystal" Bowl, 6-1/2", $35; Mayonaisse Bowl, footed, 5-1/2", $40; Vase, 7-1/4", $50; Candlesticks, 1-lite, bowl-shape, pair $65. *Courtesy of Brenda Beckett.*

Imperial #588 "Pie Crust" Bowl, 2-handled, 10", $45; Plate, 2-handled, 12-1/2", $45. *Courtesy of Brenda Beckett.*

Imperial "Crocheted Crystal" Candlesticks, 2-lite, 4-1/4", $65; Plate, 12-1/2", $45.

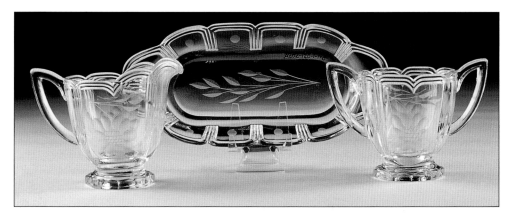

Imperial #588 "Pie Crust" Cream & Sugar, 2-3/4" and Tray, 7-1/4", set $55. *Courtesy DCMA.*

111

Indiana #607 Bowl, 11", $60; #607 Bowl, 13", $65. *Courtesy DCMA.*

New Martinsville #35 "Fancy Squares" Compote, 5-1/4", $35; Plate, 2-handled, 10-3/4", $45; Bowl, 2-handled, 10", $45. *Courtesy DCMA.*

Indiana #1011/1B4 "Tear Drop" Comport "J", 10-1/2", $55; #1011/1B1 "Tear Drop" Comport "C", 12", $65.

Indiana #1011/1A15 "Tear Drop" post-handled Sandwich, 11-1/2", $55; #1011 "Tear Drop" Comport 10", $55. *Courtesy of Brian J. Wing.*

New Martinsville #37 "Moondrops" Relish, 2-part, 10", $45. *Courtesy of Brenda Beckett.*

New Martinsville #37 "Moondrops" Cream & Sugar, regular, pair $30; individual Cream, Sugar & Tray, divided, set $50.

Viking #4500 "Janice" Bowl, flared, 10", $65; Bowl, flared, 10-3/4", $65; Bowl, turned-up, 11", $65.

Viking #4500 "Janice" Plate, 2-handled, 9", $30; #4520 Plate, 2-handled, 8", $25. *Courtesy DCMA.*

Viking #42 "Radiance" Butter & Lid, 6", $65; Cream, Sugar, 3" with Tray, 10-1/4", set $45.

Viking #42 "Radiance" Bowl, flared, 10-1/2", $65; Bowl, crimped, 12", $75; Bowl, flared, 9-1/2", $65.

Candlewick – Corn Flower

Candlewick, made by the Imperial Glass Corporation of Bellaire, Ohio, was one of the most recognized and longest running Elegant Glass patterns ever produced. For almost four decades it was a mainstay of the Hughes Company catalogues and dominated their showroom shelves. According to glass researcher Willard Kolb, in 1933 Imperial Company president Earl W. Newton brought back from New York an item in the French Cannon Ball pattern. Similar patterns with beads on the edges, such as the United States Glass Company's Atlas line and McKee Glass Company's Rays pattern, may also have inspired Imperial Candlewick designers. After a couple of years of experimentation, Imperial's craftsmen put a half dozen items into production. These initial bowls and compotes were ribbed to disguise mould lines. Then, in a brilliant stroke, they developed a uniquely designed block mould that eliminated the joint marks on items. Any signs of mould marks on the distinctive beads were easily removed during the warming-in process. What resulted was a clean, clear, uncluttered surface ideal for cuttings by Imperial and others, like W.J. Hughes.

By 1936, Candlewick was being promoted, and by the late 1930s many new items were being added to a burgeoning Candlewick catalogue. Items in this first series of Candlewick patents were under the numbers 100577 through 100579, dated 7/28/36. The first Candlewick items cut by W.J. Hughes were: 67B, the 9" low footed fruit bowl; 67D, the 10" low cake stand; 74 B, the

400/67B Compote or low Footed Fruit Bowl, ribbed, 9", $150.
Courtesy of Brenda Beckett.

4-toed, 8-1/2" bowl; 74J, the 4-toed, 7" lily bowl; and 74SC, the 4-toed 9" square, crimped bowl. It is believed that Hughes's initial order from Imperial was placed in 1939. This is substantiated in part by the absence of Candlewick in the 1938 Thornley Wrench Corn Flower photograph/catalog. By 1940, W.J. Hughes had a verbal agreement for exclusive cutting rights to Candlewick in Canada.

By the mid-1940s some two hundred and fifty different Candlewick items were available to Hughes. For the most part the uncluttered, smooth surfaces were excellent for the light cutting or grey cutting of Corn Flower, but a large number of Candlewick items were never ordered for Corn Flower cutting due to rounded surfaces, small cutting areas, awkwardness of cutting or because of prohibitive costs. On some of the smaller items, such as 400/172, the 4-1/2 inch Mint Heart, the full flower could not be accommodated so a star pattern was cut instead. As on other blanks one might also find a bud cut or a spray cut on items with space constraints. The piecrust style of decorative edge cuts, "beading," was phased out early in the 1950s. This provides an aid to dating Candlewick-Corn Flower pieces.

Candlewick Plates. Left: Corn Flower in center with wreathing effect with leaves, regular bud cut used around the edge. Right: star cut in the smaller center portion.

Two Sugars showing the Pie Crust "beading" on rim, and size variance of a half inch in diameter at the top.

In the early-1950s Hughes Corn Flower listed over one hundred Candlewick items in their first post-war sales catalog. Recently uncovered Imperial documents show that in a five-page listing Imperial offered the firm a choice

of over two hundred different Candlewick items. From an advertisement from the fall of 1954 we know that, by then, Hughes Corn Flower was the exclusive Canadian distributor for Candlewick.

The Hughes-Imperial relationship was a fruitful one for both partners, lasting for some fifty years. From the very limited information preserved from their dealings, we have evidence that in 1942 W.J. Hughes ordered $5,408.30 worth of Candlewick glass from Imperial. In October, 1952, Hughes Corn Flower purchased an entire boxcar load of Candlewick with an invoice total of $54,682.68 – a tenfold increase from 1942. The five page 1952 order consisted of in excess of 134,000 items (if one includes the individual items from sets, such as the heart-shape bonbons, ash trays, and condiment sets, the total rises to approximately 155,000 items.) There is, in fact, only one Candlewick-Corn Flower invoice that has surfaced to date – from that very same 1952 order. It has been suggested that this was the single largest purchase of Candlewick in Imperial's history.

24th October 1952 invoice from Imperial Glass for Candlewick shipment. It is no wonder that we see so many 400/96 Salt & Peppers with either the plastic or chrome lids – 720 dozen of each type were ordered! *Courtesy DCMA.*

Only two items in colored Candlewick-Corn Flower have thus far surfaced, in amber and in pink. No Viennese blue with Corn Flower cutting has yet been found. It may not exist. What has been mistakenly identified as Viennese blue cream and sugar sets are simply items from batches of crystal glass that have a faint grey/blue cast to them. Technological innovations in the glass industry after 1950 led to much truer crystal-clear glass being produced in later years. The two colored Corn Flower-Candlewick items confirmed to date are the 5-inch amber and 6-inch pink Ashtrays, 400/133 and 400/150 respectively. Since these ashtrays were sold in sets of three as 400/550, one would expect at some point the small blue 400/440 Ash tray should be found to complete the set.

Imperial Candlewick #400/133 amber Ashtray, 5", $55; #400/150 pink Ashtray, 6", $75. *Courtesy DCMA.*

Imperial, at its peak, offered some 300 items at one time in their Candlewick line. All tolled the number of items produced in the half-century run of Candlewick by Imperial astoundingly number in excess of seven hundred. Production lasted from 1936 to 1984, when the plant closed. Hughes Corn Flower was cut on Candlewick for virtually its entire production run. The longevity of their dealings was largely due to the success of that gorgeous crystal line of glassware that Imperial called Candlewick.

China and Glass Gift Buyer advertisement, September, 1958, featuring 400/20 Candlewick-Corn Flower Punch Set. *Courtesy DCMA.*

A magnificent view of Candlewick from the 1952 showroom. The front table shows off the Punch Bowl, while the top left shelf features Candlesticks and the right hand shelves showcase over a dozen Condiment Sets and Marmalades. *Courtesy DCMA.*

Showroom at 102 Tycos Drive at Opening Day 1952. The tables were stocked full of Corn Flower cut Candlewick. Visible on the closest table are Celery Dishes, Plates, and the 2-piece Tidbit. *Courtesy DCMA.*

400/118 Ashtray, $10; 400/652 Ashtray, square, $40; 400/440 Ashtray, round, $10. *Courtesy DCMA.*

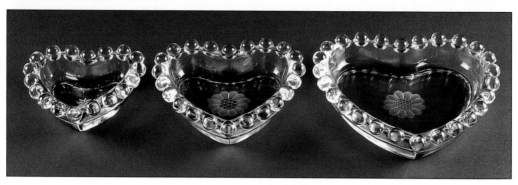

400/750 Bonbon set, 4-1/2", 5-1/2" & 6-1/2" hearts $50.

400/179 Bell, 4" $100. *Courtesy of Brian J. Wing.*

400/40/0 Basket, handled, 6-1/2", $65.

400/65 Covered Vegetable Bowl, 8". Note that the cover is not pictured. $200.

400/273 Basket, beaded handle, 5", $225.

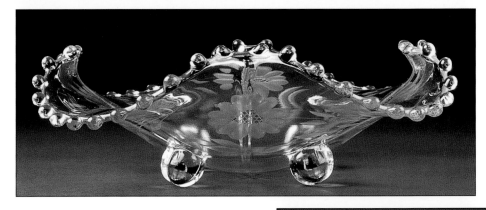

400/74SC Bowl, crimped, 4-toed, ribbed, 9", $75.

400/69B Round Vegetable Bowl, 8-1/2", $45; 400/72B Bowl, 8", 2-handled, 8", $30.

400/49H Bowl, heart, unhandled, 5", $25; 400/51H Bonbon, heart-shaped, handled, 6", $30; 400/51T Wafer Tray, center-handled, 6", $30.

400/53S Square Round Bowl, 5". Specially made for W.J. Hughes and for Lipman Sterling Limited, Toronto, $100.

400/206 Nappy, 3-toed, 4-1/2", $85.

400/75F Float Bowl, 11", $50;
400/92F Float Bowl, 12", $75.

400/92R Mushroom Center Bowl, 12", $100.
Courtesy of Brenda Beckett.

400/106B, Bowl, belled centre, graduated beads, 12", $65; 400/75 Fork & Spoon set, beaded, $70.

400/276 "California" Butter, oblong covered with beads on lid, $100; 400/144 Butter Dish & Cover, round, tab-handled, 5-1/2", $50.

400/276 "California" butter, oblong covered with no beads on lid, $125.
Courtesy of Lynda Edwards.

400/224 Candlesticks, beaded tri-stem, 5-1/2", pair $400; 400/100 Candlesticks, 2-lite, pair $70; 400/147 Candlesticks, 3-lite, 1-bead stem, pair $75; 400/80 Candlesticks, low, 3-1/2", pair $45.

120

400/40F Candlesticks, round flower, 6",
pair $85. *Courtesy DCMA.*

400/40S Candleholders, square-
shape, 6-1/2", pair $155.

400/175 Candlesticks, tall,
3-bead stem, 6-1/2", pair
$300. *Courtesy DCMA.*

400/224 Candlesticks, beaded tri-
stem, 5-1/2", pair $400.

121

400/259, Candy, shallow round covered, 7", $120.

400/105 Celery, two open handles, 13-1/2", $45.

400/134 Cigarette Box, domed cover, 5-1/4", $65. *Courtesy of Lynda Edwards.*

400/66B Compote, low, 2-bead stem, 5-1/2", $60. *Courtesy DCMA.*

400/66B Compote, low, no beads on stem, $50.

400/1769 Condiment Set,
4 piece, $95.

400/129/29 individual Cream & Sugar on 400/29 Tray, set $45.
400/29/30 Cream, Sugar & Tray Set, three versions. $40 per set.

400/31 Cream & Sugar Set, $45.

400/2696 Hospitality Set, 1 section of 6, 6-1/2".
$500 per section. *Courtesy DCMA.*

400/20 Punch Set, 6-quart Bowl, 17" Plate, 12 Cups,
Ladle, $500. *Courtesy DCMA.*

400/1989 Marmalade Set, 3-piece, $45.

Box for 400/20 Punch Set. $50. *Courtesy DCMA.*

400/89 Marmalade Set, 4-piece, $50. *Courtesy of Brian J. Wing.*

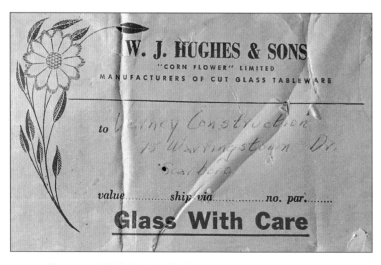

Close-up of W.J. Hughes & Son shipping
label on Punch Set. *Courtesy DCMA.*

400/1589 Twin Jam Set, $145.

124

400/23 Mayonnaise Set, 3-piece, $45;
400/42/3 Mayonnaise Set, 3-piece,
$50; 400/165 Ladles, 3-beads, $15.

400/84 Divided Bowl, 6-1/2", $40; 400/52
Divided Bowl or Jelly, handled, 6-1/2",
$35; 400/135 Ladle, 2-beads, 6-1/2",
$15; 400/165 Ladle 3-beads, 6", $15.

400/5629 Mustard & Ketchup Set, with 400/29 Tray, 7",
set $135.

400/156 Mustard, covered, with Spoon, $65.

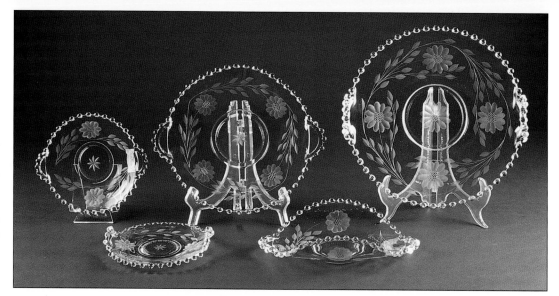

400/42D Plate, 2 open handles, 6", $17; 400/62D Plate, 2 open handles, 8", $25; 400/72E Plate, 2 turned-up handles, 10", $55; 400/42E Plate, 2 turned-up handles, 6", $25; 400/52E Plate, 2 handles, sides turned up, $35.

400/119 Oil or Vinegar with Stopper, $55.

400/124 Platter, Oval, 12", $85. *Courtesy of Brian J. Wing.*

400/58 Trays, Oval Pickle or Celery, 7-1/2", $25 each. *Left:* Corn Flower cut with horseshoe leaf pattern, edge with Corn Flower only. *Right:* Leaf pattern only in middle, combination bud and Corn Flower cut on the edge. *Courtesy of Brenda Beckett.*

400/55 Relish Tray, 4-part, handled, 8-1/2", $35; 400/56 Relish Tray, 3-part, 10-1/2", $50; 400/56 Relish Tray, 2-part, 6-1/2", $20.

400/75 Salad Set, Fork & Spoon, beaded, pair $65.

400/215 Relish Tray, oblong, 4-part, handled, 12", $85; 400/268 Relish, oval, 2-part, handled, 8" $25.

701 Salad Set, Fork & Spoon, reeded, originally used with Candlewick. Note the early Corn Flower label on the Spoon, pair $75. *Courtesy of Brenda Beckett.*

400/56 Relish, 5-part, pinwheel sections, 10-1/2", $55.

400/109 Individual Salt & Pepper Set, pair $22; 400/96 Salt & Pepper Set, pair $18; 400/96 Salt & Pepper Set, nine beads from 1940s, pair $30; 400/247 Salt & Pepper Set, straight-sided, 4", pair $30.

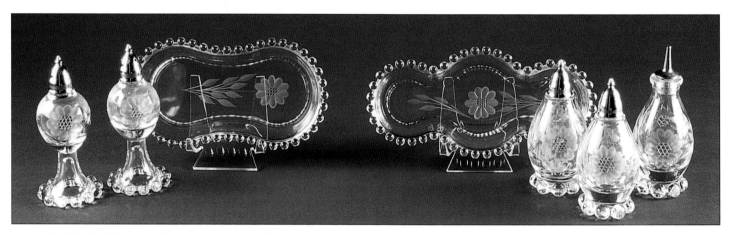

400/190 Salt & Pepper Set with 400/29 Tray, $65; 400/96 Salt & Pepper, $18; 400/117 Bitters & Metal Tube, $55; and 400/171 Tray, $30.

400/96 Salt & Pepper Set, nine beads from 1940s. With mother-of-pearl lids, possibly after-factory additions, pair $45. *Courtesy of Brenda Beckett.*

400/64 Nut Dish, 2", $15. With star cut only. 400/156 Salt Spoon, $15. *Courtesy of Lynda Edwards.*

400/88 Flat Compote, 5-1/2", $35;
400/67D Low Cake Stand, 10" $75.

400/68F Fruit Tray, handled, 10-1/2", $150.

400/221 Lemon Tray, 5-1/2", $40.

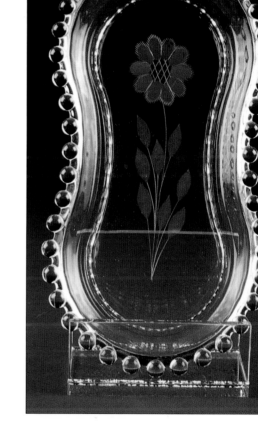

400/29 Tray, 7", $18.

400/149D Mint Tray,
handled, 9", $35; 400/68D
Pastry Tray, 12", $65.

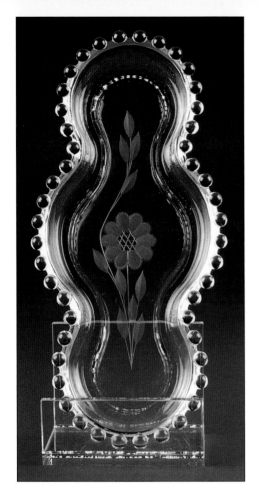

400/171 Tray, 8", $30.

400/87C Vase, crimped, 8", $75; 400/287C Vase, crimped, 6", $55. *Courtesy of Brenda Beckett.*

400/287F Fan Vase, 6", $55; 400/87F Fan Vase, 8-1/2", $75. *Courtesy of Brenda Beckett.*

400/96 Salt Shakers converted into Atomizers, $55 each. Shakers in green and violet were sold to T. Eaton Company possibly also to use for atomizers. It's a pity that the Hughes Corn Flower never was cut on these colors! *Courtesy of Brian J. Wing.*

400/143C Flip Vase, crimped top, 8" $100; 400/87C Vase, crimped, 8" $75.

400/25 Ball Bud Vase, 4", $65; 400/107 miniature Bud Vase, 5-3/4", $75.

400/2701, Tid-Bit Plate with hole, 10-1/2", $45. This handle is not an Imperial supplied one. *Courtesy of Brenda Beckett.*

1951 catalog page showing Candlewick: 400/49H Bowl, heart unhandled, 5", $25; 400/51T Wafer Tray, center-handled, 6", $30; 400/89 Marmalade Set, 4-piece, $50; 2701 Tid-Bit Set, two-tier, $85. *Courtesy DCMA.*

131

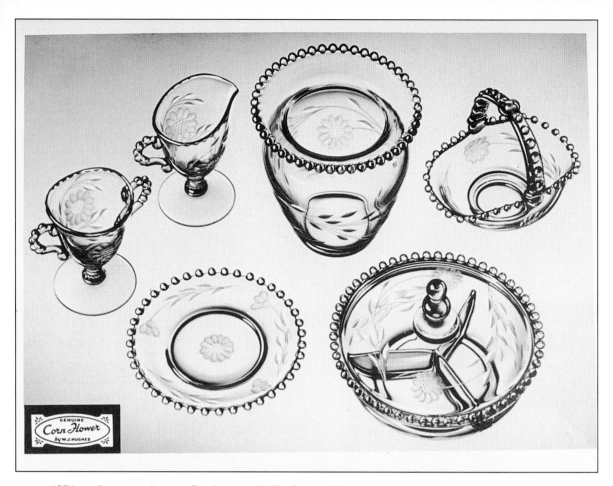

1951 catalog page showing Candlewick: 400/31 Cream & Sugar, footed set, beaded handle.
$45; 400/1D Plate, Bread & Butter, 6", $10; 400/198 the rare Vase, 6", $325; 400/260 Candy
Box & Cover, 3-part, 7", $275; 400/273 Basket, beaded handle, 5". *Courtesy DCMA.*

Retail Price List,
effective July 3rd, 1951.
Courtesy DCMA.

GENUINE "CORN FLOWER" by W. J. HUGHES   Retail price list effective July 3, 1951

CANDLEWICK			CANDLEWICK (continued)		
1/6	6" bread and butter plate	$ 2.45 ea.	10/8	8" two-handled bowl	$ 4.95 ea.
1/8	8" salad plate	3.75 ea.	10/10	10" two-handled bowl	6.95 ea.
			10/12	12" two-handled bowl	9.50 ea.
2/14	14" hostess plate	9.00 ea.			
			11/2	salad fork and spoon	3.15 set
3/17	17" torte plate	15.95 ea.	11/4/4	four-piece salad set	15.00 ea.
			11/8	8½" vegetable bowl	4.50 ea.
4/13	13" cupped-edge plate	6.75 ea.	11/10	10½" salad bowl	6.25 ea.
4/17	17" cupped-edge plate	15.95 ea.			
			12/10	10½" salad bowl	6.25 ea.
5/6	6" two-handled plate	1.85 ea.	12/12	12" centre bowl	7.75 ea.
5/7	7" two-handled plate	2.90 ea.			
5/8	8" two-handled plate	4.35 ea.	13/11	11" float bowl	6.50 ea.
5/10	10" two-handled plate	5.25 ea.			
5/12	12" two-handled plate	6.95 ea.	14/5	5" unhandled heart bowl	2.35 ea.
5/14	14" two-handled plate	9.50 ea.	14/10	10" unhandled heart bowl	6.00 ea.
6/8	8" two-handled crimped plate	4.50 ea.	15/0	punch ladle	2.55 ea.
6/10	10" two-handled crimped plate	5.75 ea.	15/4/15	fifteen-piece punch set	52.50 ea.
6/12	12" two-handled crimped plate	6.95 ea.	15/6	6 qt. bowl	13.50 ea.
6/14	14" two-handled crimped plate	9.75 ea.	15/12	punch cup	1.85 ea.
7/6	5½" centre-handled lemon tray	2.50 ea.	20/4	four-piece relish set	10.95 set
7/9	9" centre-handled mint tray	5.00 ea.	20/5	five-piece cold meat set	6.40 ea.
7/12	12" centre-handled pastry tray	7.95 ea.	20/6	6" 2 section relish	3.50 ea.
			20/7	seven-piece relish set	6.70 ea.
8/6	5½" cheese stand	2.75 ea.	20/9	8½" 4 section relish	6.25 ea.
8/10	10" cake stand	7.95 ea.	20/10	10½" 6 section relish	7.85 ea.
			20/11	10½" 5 section relish	7.75 ea.
9/2	two-tier tid-bit set	9.95 ea.	20/12	12" 4 section relish	6.95 ea.
10/5	5" two-handled bowl	1.75 ea.	21/8	8" oval pickle tray	3.65 ea.
10/6	6" two-handled bowl	2.65 ea.			
10/7	7" two-handled bowl	3.75 ea.	22/9	9" oval tray	2.95 ea.

Above prices are subject to change without notice.

Corn Flower catalog page c.1962 showing, from left to right:
Bottom row: 400/205 3-toed Bowl, 10", $175; 400/208 3-part, Relish, 3-toed, $125; 400/207, 3-toed Candleholder, 4-1/2", pair $250 ; 400/623 Mayonnaise Set, 2-piece, $100; 400/206 3-toed Nappy, 4-1/2", $85. *Courtesy DCMA.*
Middle row: Viking Glass Company items.
Top row: "Modern" stemware line by Louie Glass Company.

Corn Flower catalog page c.1962 advertising: 400/231, 232, 233 Square Bowls, 5", 6" & 7", $100, $120, $130; 400/234 Square Relish, 7", $150; 400/652 & 653 Square Ashtrays, 5" & 6", $40 each. *Courtesy DCMA.*

Corn Flower on Candlewick Look-Alikes

Czech "Boule" Bowl, 3-footed, 8", $55.

Czech "Boule" Bowl, oval, 11", $55. *Courtesy of Brenda Beckett.*

Czech "Boule" Bouillon Cup, 2-handled, and Underplate, set $50. *Courtesy DCMA.*

Czech "Boule" Cream & Sugar, 2-1/2", pair $25; Punch Cup, 2-3/4", $12; Marmalade Jar and Lid, 4", $25.

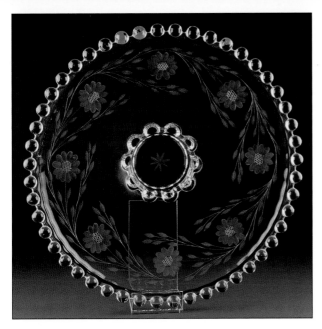

Czech "Boule" Plate, 15", $95, *Courtesy of Brian J. Wing.*

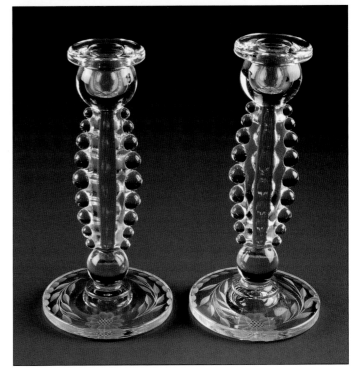

Czech "Boule" Candlesticks, 1-lite, graduated beads, 7", pair $65. *Courtesy DCMA.*

Paden City #444 "Alexander" Candlesticks, 1-lite, 4-bead stem, 6", pair $75. *Courtesy of Brenda Beckett.*

Imperial #163 Candlestick, 2-lite, graduated beads, 5". Although made by Imperial through the 1930s and 40s, they were never marketed as Candlewick, pair $85.

Czech "Boule" Plate, 10", $40. *Courtesy of Brenda Beckett.*

Czech "Boule" Trays, squared oval, $20 each.

Czech "Boule" Relish, 3-part, 8", $30;
Relish, 3-part, 11", $40.

Louie Glass Vase, footed, 1 bead stem, $55. Cut by R.G. Sherriff sometime after he left Hughes in 1939.

Listing of Candlewick-Corn Flower

The number 400 denotes the Imperial Candlewick line. The following digits or letters refer to particular items and shapes within the series (400/xyz). It is intended that this listing be as comprehensive as possible with the provision that all items have been definitively verified in collections or W.J. Hughes catalogues. Purportedly, there were some of the 3400 line stems cut for a special order in the 1950s, but none have been seen, hence these potential items are not included. It is believed that the total number of distinct Candlewick-Corn Flower items could number over two hundred. We are hopeful that the publication of this book will fill in some gaps in this inventory. The search and research continues.

Candlewick-
Corn Flower Listings

Item #400/	Shape	Size	Description	Years	Price
450	Ash Tray Set	4/5/6-1/2"	3 pc nested	1943-68, 77-79	Set $30
651	Ash Tray	3-1/4"	square	1957-67	$40
118	Ash Tray Set	4-3/8"	4 pc bridge [coaster]	1950-70	Set $40
440	Ash Tray	4"	round	1943-68, 77-79	$10
134/1	Ash Tray	4-1/2"	oblong [pin tray]	1943-63	$10
652	Ash Tray	4-1/2"	square	1957-67	$40
133	Ash Tray	5"	round	1941-67, 77-79	$8
133	Ash Tray	5"	round amber	1944-60	$55
653	Ash Tray	5-3/4"	square	1957-67	$40
150	Ash Tray	6"	round	1943-68, 77-79	$10
150	Ash Tray	6"	round pink	1944-60	$75
273	Basket	5"	beaded handle	1953-63	$225
40/0	Basket	6-1/2"	plain handled	1941	$65
179	Bell	4"	table	1949-51	$100
117	Bitters	4"	with 1" tall metal tube, 4 oz	1948-54	$65
750	Bonbon Set	4-1/2"; 5-1/2"; 6-1/2"	heart shape (173, 174, 175)	1948-63	$55
172	Heart	4-1/2"	mint, nut, bonbon, ashtray	1948-63	$12

Item #400/	Shape	Size	Description	Years	Price
173	Heart	5-1/2"	nut	1948-63	$20
174	Heart	6-1/2"	bonbon	1948-63	$23
40H	Bonbon	5"	handled-heart	1939-84	$35
51H	Bonbon	6"	handled-heart bonbon	1943-68	$30
206	Bowl	4-1/2"	3-toed nappy	1960-67	$85
49H	Bowl	5"	heart unhandled	1939-53	$25
231	Bowl	5"	square nappy	1957-63	$100
53S	Bowl	5"	round-square, square base	1954	$100
42B	Bowl	5"	nappy or fruit, 2-handled	1937-78	$15
52B	Bowl	6"	nappy, 2-handled	1937-73	$20
52	Bowl	6"	jelly, 2-handled divided	1950-76	$30
232	Bowl	6"	square nappy	1957-63	$120
85	Bowl	6-1/2"	cottage cheese	1951-73, 77-84	$45
233	Bowl	7"	square nappy	1957-63	$130
62B	Bowl	7"	nappy, 1-handled	1937-84	$25
65	Bowl	8"	covered vegetable	1941-49	$200
72B	Bowl	8"	2-handled	1937-76	$30
72	Bowl	8-1/2"	2-handled divided jelly	1953-63	$100
69B	Bowl	8-1/2"	round vegetable	1951-72, 75-84	$45
74B	Bowl	8-1/2"	round, 4-toed, ribbed	1937-41	$75
74SC	Bowl	9"	crimped, 4-toed, ribbed	1937-46	$75
49H	Bowl	9"	heart unhandled fruit bowl	1939-55	$155
67B	Bowl	9"	low footed fruit, ribbed	1937-39	$150
205	Bowl	10"	3-toed bowl	1960-67	$175
145B	Bowl	10"	2-handled	1943-84	$60
63B	Bowl	10-1/2"	belled, graduated beads	1939-70	$65
75B	Bowl	10-1/2"	salad	1939-67	$50

Item #400/	Shape	Size	Description	Years	Price
75F	Bowl	11"	float, cupped edge	1939-75	$50
92F	Bowl	12"	float	1943-67	$75
113B	Bowl	12"	round, 2-handled	1941-55	$150
106B	Bowl	12"	center, belled, graduated beads	1939-61	$65
92R	Bowl	13"	mushroom center bowl	1941	$100
20B	Bowl	13"	6 quart punch	1939-79	$200
144	Butter	5-1/2"	covered, tab-handled	1943-67, 76-80	$50
276	Butter	6-3/4"	oblong covered "California", beads on lid	1962-68	$100
276	Butter	6-3/4"	oblong covered "California", no beads on lid	1950-61	$125
67D	Cake Stand	10"	1 bead stem	1941-67	$75
80	Candlesticks	3-1/2"	low	1937-76	pair $45
207	Candlesticks	4-1/2"	3-toed	1960-67	pair $250
129R	Candlesticks	4-1/2"	urn-shape	1943	pair $200
224	Candlesticks	5-1/2"	beaded tri-stem	1950-55	pair $400
40F	Candlesticks	6"	flower, round	1950-60	pair $85
175	Candlesticks	6-1/2"	tall, 3-bead stem	1948-55	pair $300
40S	Candlesticks	6-1/2"	blown square shape	1950-51	pair $155
100	Candlesticks	4-1/4"	2-lite	1939-68	pair $70
147	Candlesticks	5-1/2"	3-lite, 1-bead stem	1943-65	pair $75
110	Candy	7"	covered, 3-part round	1941-63	$100
259	Candy	7"	covered, shallow round	1949-70	$120
260	Candy Box	7"	covered	1953-60	$275
105	Celery	13-1/2"	2 open handles	1939-67	$45
134	Cigarette box	5-1/4"	domed cover	1941-67	$65
88	Compote	5-1/2"	cheese stand, 1 bead stem	1949-67	$35
66B	Comport	5-1/2"	low, 2-bead stem	1950-80	$60
66B	Comport	5-1/2"	low, no beads on stem	1937-1949	$50

Item #400/	Shape	Size	Description	Years	Price
45	Comport	5-1/2"	4 bead stem	1941-67	$50
67C	Comport	9"	crimped	1953-55	$195
1769	Condiment Set, 4 pc.		tray, oil, salt & pepper (96, 119, 171)	1948-55	set $95
1574	Condiment Set, 5 pc.		9-1/4" oval tray, 4 oz. oil, salt & Pepper (159, 164, 167)	1948-51	set $150
5996	Condiment Set, 5 pc.		bottle, salt and pepper, relish, tray (89, 96, 119, 159)	1948-50	set $135
1786	Condiment Set, 6 pc.		relish, salt & pepper, tray (89, 96, 171)	1948-60	set $100
122	Cream and Sugar	2-1/2"	individual	1949-80	set $30
30	Cream and Sugar	3-1/4"		1941-84	set $20
31	Cream and Sugar	4-1/2"	footed, beaded handle	1939-67	set $45
122	Cream and Sugar	2-1/2"		1949-80	set $25
75	Fork and Spoon	9-1/2"; 9-1/4"	beaded	1939-67	set $65
701	Fork and Spoon	9-1/2"; 9-1/4"	reeded, sold in early years with Candlewick	1932-68	set $75
8918	Marmalade Set, 3 pc.		footed old fashion, lid, ladle (18, 89, 130)	1949-63	set $120
1989	Marmalade Set, 3 pc.		lid, spoon, bowl (old fashion) (89, 130, 131)	1941-48	set $45
89/3	Marmalade Set, 3 pc.		bowl, lid, ladle (35, 89)	1956-59	set $60
89	Marmalade Set, 4 pc.		saucer, bowl, lid, ladle (130, 35, 89)	1937-43	set $55
623	Mayonnaise Set, 2 pc.		6", 3-toed bowl; 5" ladle (183, 165)	1960-66	set $100
40	Mayonnaise Set, 3 pc.		5-1/2" bowl; 7" plate; 5" ladle; no beads(40, 40D, 615)	1937-41	set $40
42/3	Mayonnaise Set, 3 pc.		4-1/4" bowl, 5-1/2" plate, ladle (42B, 42D, 130)	1948-1970	set $40
23	Mayonnaise Set, 3 pc.		5-1/4" bowl, 7" plate, 5" ladle (23B, 23D, 165)	1939-69	set $45
84	Mayonnaise Set, 4 pc.		6-1/2" divided bowl, 8" plate & 2 ladles (84, 84D, 65)	1938-60	set $75
156	Mustard		low-footed, 2-bead finial	1943-67	$65
5629	Mustard & Catsup Set, 3 pc		7" tray, 2 covered mustard, 2 spoons (29, 156)	1948-50	set $135
64	Nut Dish	2"	also called nut dip; nut cup; sugar dip; ash tray	1941-67	$15

Item #400/	Shape	Size	Description	Years	Price
177	Oil or Vinegar	6"	with stopper, 4 oz	1948-55	$65
275	Oil Cruet	7-1/2"	with stopper, 6 oz	1950-60	$55
2911	Oil and Vinegar on Tray, 3 pc.		7" tray, 2 oil or vinegar, 6 oz.(29, 119)	1946-50	set $95
1D	Plate	6"	bread & butter	1937-77	$10
42D	Plate	6"	two open handles	1937-73	$17
42E	Plate	6"	turned-up handles	1937-43	$25
52D	Plate	7"	two-handled	1937-73	$25
52E	Plate	7"	sides turned up	1953-58	$35
62C	Plate	8"	2-handled crimped	1950-68	$40
5D	Plate	8"	salad/luncheon	1937-84	$15
62D	Plate	8"	two-handled	1937-68	$25
72C	Plate or Tray	10"	2-handled crimped	1949-73	$45
72D	Plate	10"	2-handled	1937-76	$45
72E	Plate	10"	turned-up handles	1937-43	$55
145C	Plate or Tray	12"	2-handled crimped	1949-68	$50
145D	Plate	12"	two open handles	1943-84	$45
75V	Plate	13"	cupped edge torte	1939-84	$45
92D	Plate	14"	hostess	1938-67	$65
113C	Plate	14"	2-handled crimped	1949-53	$175
113D	Plate	14"	two-handled	1941-67	$65
20D	Plate	17"	flat edge torte	1939-73	$65
20V	Plate	17"	torte or sandwich, cupped edge	1949-71	$75
124	Platter	12"	oval	1948-67	$85
37	Punch Cup	2-3/4"	7 oz	1938-84	$20
91	Punch Ladle	13"	1 spout, large	1939-82	$40
20	Punch Set		6-quart bowl, 17" plate, 12 cups, ladle (20B, 20V, 37, 91)	1939-79	$500

Item #400/	Shape	Size	Description	Years	Price
54	Relish	6-1/2"	2-section, 2-handled	1937-71	$20
234	Relish	7"	square, 2-section	1957-63	$150
268	Relish	8"	2-section oval	1963-84	$30
55	Relish	8-1/2"	4-section, 4-handled	1937-84	$35
208	Relish	10"	3-section, 3-toed	1960-70	$125
112	Relish	10-1/2"	5-section	1941-46	$55
56	Relish	10-1/2"	5-section, 5 handled	1949-79	$55
215	Relish	12"	4-section oblong handled	1949-61	$85
209	Relish	13"	5-section	1959-61	$225
1112	Relish and Dressing, 4 pc.		10-1/2" 5-part relish with 5 tab handles, marmalade with lid, 5" 3-bead ladle (112, 89, 130)	1941-63	set $120
925	Salad Set, 4 pc.		12" bowl, 13-1/2" flat plate, fork/spoon (92B, 92V, 75)	1948-51	set $150
96	Salt and Pepper Set	3-1/2"	chrome/plastic tops	1944-84	pair $18
109	Salt and Pepper Set	5"	individual	1941-73	pair $22
167	Salt and Pepper Set	4-1/2"	high, chrome tops	1948-70	pair $30
190	Salt and Pepper Set	3-1/2"	footed, chrome tops	1948-67	pair $55
247	Salt and Pepper Set	4"	straight sided	1949-84	pair $30
169	Sauce Boat & Plate Set	8"	oval plate	1948-67	pair $175
169	Sauce Boat	8"	no beads	1948-67	$120
169	Sauce Plate	8"	oval	1948-67	$75
2701	Tid-bit Set		two-tier, 7-1/2" plate, 10-1/2" plate (270, 271)	1950-67, 1977-78	$85
96T	Tray	5"	oblong	1948-68	$20
221	Tray	5-1/2"	center-handled lemon, tri-stem	1950-67	$40
51T	Tray	6"	heart handled wafer	1944-73	$30
111	Tray	6-1/2"	with 2 divisions	1941-49	$90

Item #400/	Shape	Size	Description	Years	Price
269	Tray	6-1/2"	individual server, wedge shaped	1961-63	$500
52E	Tray	7"	2-handled muffin	1937-41	$35
29	Tray	7"	oblong tray for cream & sugar	1941-84	$18
57	Tray	7-1/2"	oval pickle	1937-43	$35
171	Tray	8"	for condiment sets 1769 & 1786	1948-60	$30
58	Tray	8-1/2"	oval pickle	1937-76	$25
149D	Tray	9"	center-handled mint	1943-67	$35
159	Tray	9"	oval	1943-67	$30
217	Tray	10"	pickle, two-handled oval	1950-67	$45
68F	Tray	10-1/2"	center-handled fruit	1939-43	$150
68D	Tray	12"	pastry, center heart handled	1939-77	$65
142	Tumbler	3-1/4"	cocktail, 3-1/2 oz	1937-43	$15
1589	Twin Jam Set, 3 pc.		9-1/4" oval tray, 2 marmalade bowls & covers, 2 ladles(159, 89)	1946-60	set $145
25	Vase	4"	footed ball bud	1939-60	$75
40V	Vase	4-1/2"	miniature	1950-55	$75
107	Vase	5-3/4"	footed bud	1950-67	$75
198	Vase	6"	rose bowl	1953-54	$325
287C	Vase	6"	crimped	1953-63	$75
287F	Vase	6"	fan	1953-63	$55
74J	Vase	7"	lily bowl, 4-toed, ribbed	1937-41	$225
87C	Vase	8"	crimped	1951-63	$75
143C	Vase	8"	crimped flip	1943-55	$100
87F	Vase	8"	fan	1939-41	$75
119	Vinegar	7-1/2"	with stopper, 6 oz	1941-50	$55

Part V

The 1938 Advertising Photographs

By far the most informative source of information about glass blanks purchased by W.J. Hughes through the 1930s is found in a series of twenty-four advertising photographs. Although they have been referred to as a "catalogue," what currently exist are simply twenty-four loose pages of photographs mounted on linen-backed cardboard. There is no indication that these pages were ever bound. On the borders of many of these unnumbered pages, and on some of the photos themselves, are notations about item name, line number, size, cost, color, height, and capacity. These notations were not systematically done and proved to be of relatively little value for research. The Toronto-based photographer, J.

Thornley Wrench, produced the photographs for this "catalog" in the late 1930s. We can ascertain that the catalogue dates no earlier than 1938 since that was the first year of production of Fostoria's Coronet line, several items of which are found in these photographs. Companies identified as being represented in these pages are Beaumont, Cambridge, Duncan Miller, Fostoria, Heisey, Lancaster, New Martinsville, and Paden City. It is believed that these images were used in W.J. Hughes's showroom and for his first distributor, Haddy, Body and Company. This Toronto-based firm had a team of commissioned salesmen representing various lines of giftware and jewelry. The Haddy, Body territory covered all of Ontario.

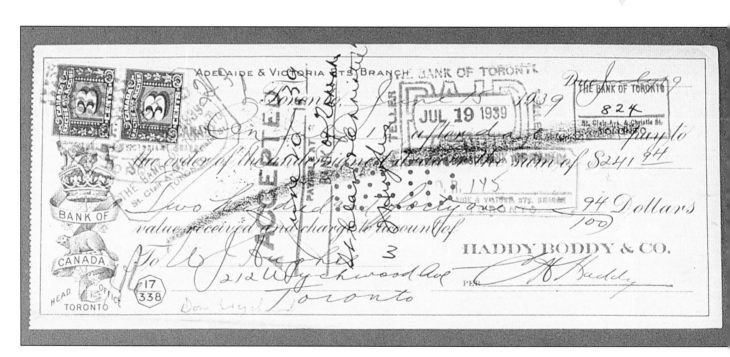

19 July 1939 check from Haddy, Boddy & Co. to W.J. Hughes.

Top row, left to right: Heisey #1229 "Octagon" Cheese Dish, 6"; Salad Plate, 7-1/2"; Bread & Butter Plate, 6". *Bottom row, left to right:* Tiffin #04 Bell Tumbler, 4-1/4"; Tiffin #114, 3 pint Jug; Tiffin #02 Table Tumbler, 3-3/4"; Heisey #173 Tankard, 54 oz.

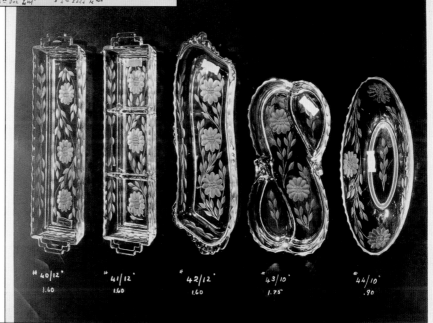

Left to right: Heisey #4044 "New Era" Celery Tray, 13"; Heisey #4044 "New Era" Relish, 3 compartments, 13"; Heisey #135 "Empress" Celery Tray, 13"; Heisey #135 "Empress" Combination Relish, 3-part, 10"; Heisey Celery Tray, 10".

Top row, left to right: Heisey Footed Salver, low, 10"; Heisey #1186 "Yeoman" individual Ashtray; Toothpick Holder; Heisey #1184 "Yeoman" Coaster; Heisey #4159 "Classic" Jug; Heisey #461 "Banded Picket" Basket, 9-1/2". *Bottom row, left to right:* Heisey #1184 "Yeoman" Floral Bowls, 12", 14", 20".

145

Top row, left to right: Heisey #1226 "Octagon" Mayonnaise, footed, 5-1/2"; Heisey Celery Tray; Tiffin #348 Candlestick, 4";
Bottom row, left to right: Paden City Syrup & metal Lid, 4-1/4"; Cambridge #6004 Vases, footed 6" & 8"; Cambridge #274 Bud Vase; Cambridge #272 Bud Vase; Heisey #21 "Aristocrat" Candlesticks, 1-lite, 7".

Top row, left to right: Lancaster #879 Cream & Sugar; Tiffin #6 Cream & Sugar.
Bottom row, left to right: Tiffin #14185 Cream & Sugar; Tiffin #3 Cream & Sugar.

Top row, left to right: Heisey #1184 "Yeoman" Round Lemon Dish; Heisey #4122 "Weaver" Marmalade and Lid; Heisey Toothpick Holder, footed; Heisey #1184 "Yeoman" Lemon Dish & Cover, 5".
Bottom row, left to right: Tiffin #5831 Candelabrum, 2-lite; Tiffin #348, Centrepiece Bowl, 3-footed, 11".

Top row, left to right: Paden City Crow's Foot #412-1/2 Candy Box, 3-sections; Paden City #417 Bowl, crimped.
Bottom row, left to right: Paden City #412 low Footed Compote, flared & crimped, 7-3/4"; Paden City #412 low Footed Cake Plate.

Top row, left to right: Lancaster #868/9 Bowl, 4-footed, 10-1/2"; Lancaster #T1377 Crimped Bowl, 6"; Lancaster #767/1 Crimped Bowl, 3-footed, 6-1/2".
Bottom row, left to right: Lancaster #869/1 Bowl, 4-footed, 9"; Tiffin Decanter & Stopper; Tiffin #020 Table Tumbler; Lancaster #R767/6 Rose Bowl, 6", $155.

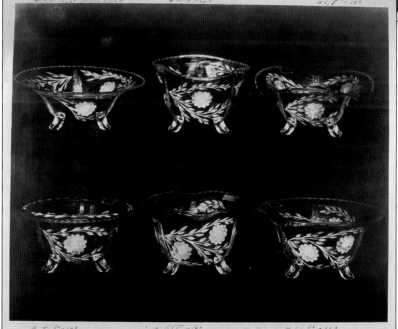

Series of Beaumont Bowls, rolled & crimped, 3-footed.

Top row, left to right: New Martinsville #4222 "Radiance" Relish, 2-compartment, 7"; Relish, 4-part; Fostoria #2560 "Coronet" Tid-Bit, flat, 3-toed, 8-1/4"; Heisey Lemon Dish & Cover; *Middle row, left to right:* Heisey #1495 "Fern" Mint, handled, 6"; Rose Bowl with Flower Frog Cover; Heisey Bowl; Heisey #1495 "Fern" Whipped Cream or Mayonnaise, 1-handled, 5". *Bottom row, left to right:* Fostoria #2560 "Coronet" Relish, 2-part, 2-handled, 6-1/2"; Jelly, 6"; Fostoria #2496 "Baroque" Relish, 2-part; Fostoria #2560 "Coronet" Bowl, 3-footed, 7".

Top row, left to right: Fostoria #2375 "Fairfax" Sweetmeat, 2 handles, 6"; Fostoria #2375 "Fairfax" Lemon Dish, 2 handles, 6-3/4"; Fostoria #2375 "Fairfax" Whipped Cream, 2 handles, 5-1/2". *Bottom row, left to right:* New Martinsville #42 "Radiance" Nut Dish, 5"; Fostoria #2510 "Sunray" Relish, 4-part, 8"; New Martinsville #42 "Radiance" Nut Dish, 5".

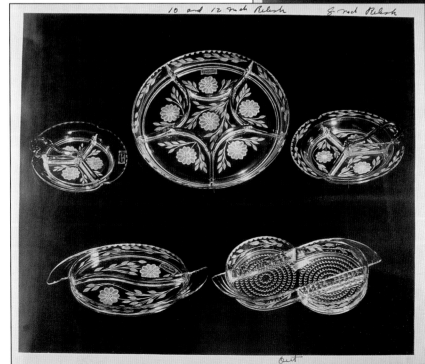

Series of Duncan Miller Relishes.

148

Top row, left to right: Lancaster 822/655 Cheese & Cracker Set; Picture Frame with fruit & leaf cut; Lancaster 895/1 Handled Fruit, 9".
Bottom row, left to right: Lancaster #833 Candlestick; Lancaster Console Bowl; Heisey #1229 "Octagon" Hors' D'Oeuvre, 13"; Heisey #1229 "Octagon" oval Dessert Dish, 8".

Top row, left to right: Heisey #1229 "Octagon" Muffin Plate, 10"; Tray, 4-footed; Relish, 2-part.
Bottom row, left to right: New Martinsville #35 "Fancy Squares" centre-handled Sandwich; Heisey #1206 Floral Bowl, with flower frog, 12".

Series of nine Divided Relishes including Paden City.

Top row, left to right: Beaumont Bowl, 3-footed, crimped; Tumbler, Iced Tea; New Martinsville #37 "Moondrops" Cream & Sugar.
Bottom row, left to right: Beaumont Bowl, 3-footed; New Martinsville #37 "Moondrops" Tumbler, 4 oz.; New Martinsville #37 "Moondrops" Cocktail Shaker; New Martinsville #35 "Fancy Squares" Mayonnaise Set.

Top row, left to right: New Martinsville #37 "Moondrops" Round Bowl, 3-footed; New Martinsville #37 "Moondrops" Crimped Bowl, 3-footed.
Bottom row, left to right: Tiffin #8896 Relish, 3-part, 9-3/4"; New Martinsville #37/4 Relish, 3-part, 2-handled, 12-3/4"; Tiffin #8897 Relish, 3-part, 9-3/4".

Left to right: New Martinsville #4222 "Radiance" Relish, 2-compartment, 7"; New Martinsville #4226 Relish, 3-compartment, 8"; New Martinsville #4228 Relish, 3-compartment, 3-handled, 8"; New Martinsville #42 Plate, 11"; Salad Plate, 8-1/2"; Sherbet Plate, 6-1/2".

Two Plates, 14" and 18".

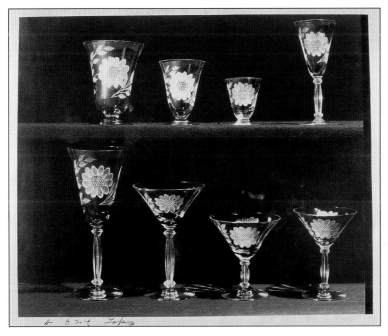

Top row, left to right: Tiffin #020 Table Tumbler; Seltzer; Whiskey; Tiffin #15024 Wine.
Bottom row, left to right: Tiffin 024 Goblet, 9 oz.; Saucer Champagne, oz.; Sundae; Cocktail, 3-1/2 oz.

Top row, left to right: Tiffin #153 Cocktail; Wine, 3 oz.; Cordial; *Bottom row, left to right:* Tiffin #153 Goblet, 10 oz.; Saucer Champagne, 6 oz.; Sundae.

151

Top row, left to right: Tiffin #020: Table Tumbler; Seltzer or Oyster Cocktail; Tiffin #14185 Whiskey; Tiffin #14196: Claret; Wine, 2 oz.. *Bottom row, left to right:* Tiffin #14196: Goblet; Saucer Champagne; Sundae; Cocktail.

Top row, left to right: crystal with rose trim Tiffin #020 Table Tumbler; Tiffin #14185 Seltzer; Whiskey; Tumbler, 9 facets around bottom, 10 oz.; Tumbler, 8 oz. *Bottom row, left to right:* Tiffin #14196: Goblet; Saucer Champagne; Sundae; crystal with green trim Cocktail, 2-1/2 oz.; Claret; Wine, 2 oz.

Top row, left to right: W.J. Hughes's "Royal" cut stemware: Ice Tea; Wine, Claret, Cocktail; Sundae, Saucer Champagne with Liner; Goblet. *Bottom row, left to right:* "Corn Flower" cut Ice Tea; Wine, Claret, Cocktail; Sundae, Saucer Champagne; Goblet.

Canadian Depression Glass Association

The Canadian Depression Glass Association provides information about Depression Glass and related topics to fellow "Preservationists of Depression Era Glassware". As well the CDGA provides a marketplace for the sale of and the search for DG. The voice of the CDGA is the *Canadian Depression Glass Review*, published six times a year. Established in 1976, the CDGA has members from coast to coast in Canada, and in the United States. Originally the CDGA was formed in Western Canada as the Prairie Depression Glass Club (by mail). We encourage all members to contribute to the advancement of Depression Glass knowledge.

Memberships: $15 US/$20 CDN per year or $40 US/$50 CDN for 3 years.

Checks or Money Orders payable to the Canadian Depression Glass Association

CDGA
P.O. Box 41564
HLRPO, 230 Sandalwood Pkwy
Brampton, Ontario
L6Z 4R1 CANADA
(905) 846-2835
www.waltztime.com/CDGA

Toronto Depression Glass Show & Sale

Twice Yearly in May & November
Sunday from 10 am to 4 pm

Adam to Windsor, Corn Flower, Elegant Glass, Fire-King, Kitchen Glass & More
Glass Displays & Free Identification Clinics

Location:
German Canadian Club Hansa
6650 Hurontario Street
Mississauga, Ontario
L5W 1N3 CANADA

Sponsored by Waltz Time Antiques
For further information contact:
(905) 846-2835
walt@waltztime.com
www.waltztime.com

Dufferin County Museum & Archives

The Dufferin County Museum & Archives strives to collect, preserve, and interpret the history of Dufferin County. Opened to the public in October, 1994, the DCMA is unique among Canadian community museums. Built in the style of an Ontario bank barn, the museum's striking architecture reflects the agricultural heritage of this central Ontario region.

The DCMA collects artifacts and archival material that relates to the historical development of Dufferin County. Our collection has thousands of archival documents and artifacts–including Canadian glass and ceramics (notably Corn Flower cut glassware), furniture, wagons, machinery, agricultural implements, clothing, quilts, archival documents, and photographs.

The DCMA itself comprises over 25 000 square feet, on four floors of exhibit and storage space. The DCMA has the largest public collection of Corn Flower. It is also the home of the annual Corn Flower Festival each June. In addition the DCMA publishes of the "Corn Flower Chronicle" The Newsletter for Corn Flower Enthusiasts.

Location:
The DCMA is located at the intersection of Airport Road and Highway # 89, between Shelburne and Alliston, 100 km, approximately one hour north of Toronto.

Web Site:
www.dufferinmuseum.com

Mailing Address:
P.O. Box 120
Rosemont, Ontario
L0N 1R0 CANADA

Telephone:
Toll Free 1.877.941.7787
or 1.705.435.1881

Plan a visit to the Dufferin County Museum & Archives Enjoy one of Ontario's best community museums. For more information, please contact us at info@dufferinmuseum.com

Hughes v Sherriff

According to Ontario Court Records [1950], reported under the authority of the Law Society of Upper Canada, Hughes employed R.G. Sherriff from 1918 until 1939. His entire twenty-two years with W.J. Hughes was occupied almost entirely with cutting the Corn Flower pattern. Towards the end of 1939, Sherriff left to set up his own glass cutting business. The parting was not amicable. Upon leaving, Sherriff remarked to Hughes: "You cannot supply the demand for your product." Hughes's reply was a curt "That's my affair".

On January 7, 1940, W.J. Hughes filed for a word trademark for "Corn Flower" under provisions of The Unfair Competition Act of 1932. Within days of hearing that Sherriff was cutting, or proposing to cut, the Corn Flower pattern Hughes's lawyer fired off a terse letter. On January 10, 1940 Sherriff received this warning "against cutting that pattern [Corn Flower] or any imitation of it." No further actions were taken for almost ten years.

In the interim Sherriff was cutting a pattern that appears indistinguishable from Corn Flower in the early 1940s. By the mid-1940s he was cutting what he called Cosmos, another very similar pattern. Cosmos has a series of lines for the petals unlike Corn Flower. The Cosmos label is easily distinguished from Hughes's Corn Flower label. It was not until 1943 that he adopted a new cut pattern that he sold under the name of Crystalware.

In February of 1948 W.J. Hughes took steps to protect his alleged rights. His new solicitor wrote to the defendant and to many wholesalers and retailers and threatened to institute legal proceedings unless, among other things, the usual undertaking to desist from continuing the acts com-

R.G. Sherriff
re-worked "Crystalware"
pattern.

plained of was supplied within a week's time. Hughes also applied, on March 6, 1948, for a design mark to protect his Corn Flower label.

Legal action was commenced on 12 March 1948. The plaintiff [Hughes] asked the Court:

> to declare that the words "Corn Flower" have acquired a secondary meaning in Canada as denoting the plaintiff's wares; he also claims damages or an accounting of profits and two injunctions: first, an injunction to restrain the defendant from using his present label, and, second, to restrain him from "infringing the Plaintiff's trade marks and trade names by the use of the word 'Crystalware' …and from passing off the Defendant's products as the products of the Plaintiff". He also asks this Court to order delivery up, for destruction, of the defendant's labels and the plates or dies used in their printing.

The Honourable Arthur Mahony LeBel heard the case, Hughes v Sherriff, in the Supreme Court of Ontario on the 11th and 12th of October 1949. It was not the similarity of the cut pattern that was in question in 1949 (an editorial note in the "Canadian Patent Reporter", 1950, suggested that the late registration of the patent would have posed an "interesting question"). In the final analysis it was the label Sherriff adopted for Cosmos that the case centered on. R.G. Sherriff's label was quite close in color and in shape to the Hughes label but with "R.G. Sherriff, Toronto" instead of "W.J. Hughes" at the bottom of the label. In the judge's decision, this was deemed to be a deceptive looking label "on grounds of passing off." Hughes adopted use of his label in 1933. It must be noted that during the case it was stated that: "The pattern cut by the defendant in his glassware looks identical to that of the plaintiff. The flowers appear to be the same." Also that "He [Sherriff] chose to cut a flower design with which to adorn his wares, strikingly similar to that in use by the respondent for many years; he has adopted the respondent's distinctive form, design and coloring of label."

W.J. Hughes "Corn Flower" label.

R.G. Sherriff "Crystalware" label.

The judge stated: "Adopting a suggestion I made at that time the parties tried to effect a settlement of their differences, but they did not succeed, and I have been asked to deliver judgment." In the February 9th, 1950 judgment, Sherriff was ordered to stop using his current Crystalware label or similar one, and to destroy such labels and plates for their production. The probability of confusion and infringement with Sherriff's labels was a case of "passing off". The judge ruled "a case has not been made out for an injunction restraining the defendant from using the word "Crystalware" in connection with the sale of his glassware". In closing the judge declared that: "the plaintiff must be assumed to have suffered damage as a result of the defendant's conduct, and he is entitled to a reference to the Master of this Court to fix his damages, or for an accounting of profits, as he elects, and to costs." This was potentially a very costly ruling against Sherriff.

An appeal was heard on June 15, 1950. The court agreed with the earlier ruling about the "copying or imitating the get-up of the respondent's [Hughes's] wares" primarily because Sherriff did not "even attempt an explanation". The injunction against using the labels stood. However, the court further ruled:

"But no case has been made out for the award of other than nominal damages; the action is really quia timet [of future consequence] and the long delay in bringing the action is not explained satisfactorily. That portion of the judgment dealing with damages and directing a reference should be deleted, and in place thereof the judgment should award respondent the sum of one dollar as nominal damages, together with costs of the action."

Louie Glass Vase cut by R.G. Sherriff, 5" c.1943.

Glass Clubs

National Cambridge Glass Collectors
P.O. Box 416
Cambridge, Ohio
43725 USA
Web Site: www.cambridgeglass.org

The National Duncan Glass Society, Inc.
525 Jefferson Avenue
Washington, Pennsylvania
15301 USA
Web Site: www.duncan-glass.com

Fostoria Glass Collectors, Inc.
P.O. Box 1625
Orange, CA 92856 USA
Web Site: www.fostoriacollectors.org

Fostoria Glass Society of America
P.O. Box 826
Moundsville, WV 26041 USA
Web Site: www.fostoriaglass.org

Heisey Collectors of America, Inc.
Museum: 169 West Church Street
Newark, Ohio 43055 USA
National Heisey Glass Museum
Web Site: www.heiseymuseum.org

National Imperial Glass Collectors Society
P.O. Box 534
Bellaire, Ohio
43906 USA
Web Site: www.imperialglass.org

Paden City Glass Collectors Guild
42 Aldine Road
Parsippany, NJ 07054 USA
Web Site: pcguild1@yahoo.com

Paden City Glass Society
Box 139
Paden City, WV 26159 USA
Web Site: pcglasssociety@mailcity.com

Tiffin Glass Collectors Club
P.O. Box 554
Tiffin, Ohio
44883 USA
Web Site: www.tiffinglass.org

Westmoreland Glass Collectors Club
P.O. Box 100
Grapeville, Pennsylvania
15634 USA
Web Site: www.westmorelandglassclubs.org

Westmoreland Glass Society
P.O. Box 2883
Iowa City, Iowa
52244-2883 USA
Web Site: www.westmorelandglassclubs.org

West Virginia Museum of American Glass
P.O. Box 574
Weston, West Virginia
26452 USA
Web Site: www.allaboutglass.org

Bibliography

Bickenheuser, Fred. *Tiffin Glass Masters*. Grove City, Ohio: Glass Masters Publications, 1979.

Bickenheuser, Fred. *Tiffin Glass Masters Volume II*. Grove City, Ohio: Glass Masters Publications, 1981.

Bredehoft, Neila & Tom. *Heisey Glass 1896-1957*. Paducah, Kentucky: Collector Books, 2001.

Bredehoft, Neila. *Heisey Glass 1925-1938*. Paducah, Kentucky: Collector Books, 1986.

Dufferin County Museum & Archives Collection.

Florence, Gene. *Collectible Glassware from the 40s 50s 60s*, 7th edition. Paducah, Kentucky: Collector Books, 2004.

Florence, Gene. *Collector's Encyclopedia of Depression Glass*, 16th edition. Paducah, Kentucky: Collector Books, 2004.

Florence, Gene. *Elegant Glassware of the Depression Era*, 10th edition. Paducah, Kentucky: Collector Books, 2003.

Garrison, Myrna and Bob. *Candlewick: Colored and Decorated Lines*. Atglen, Pennsylvania: Schiffer Publications, 2003.

Garrison, Myrna and Bob. *Candlewick: The Crystal Line*. Atglen, Pennsylvania: Schiffer Publications, 2003.

Goshe, Ed. Ruth Hemminger & Leslie Pina. *Tiffin Depression Era Stems & Tableware*. Atglen, Pennsylvania: Schiffer Publications, 1998.

Harvey, Alan Burnside, editor. The Ontario Reports. Toronto, Canada: Carswell, 1950.

Long, Milbra and Seate, Emily. *Fostoria Tableware, 1924-1943*. Paducah, Kentucky: Collector Books, 1999.

Kovar, Lorraine. Westmoreland Glass, Volume 3, 1888-1940. Marietta, Ohio: Glass Press, 1997.

Mauzy, Barbara and Jim. *Mauzy's Depression Glass*, 3rd ed. Atglen, Pennsylvania: Schiffer Publications, 2004.

Measell, James. *New Martinsville Glass, 1900-1944*. Marietta, Ohio: Antique Publications, 1994.

Measell, James, ed. *Imperial Glass Encyclopedia, Vol. I*. National Imperial Glass Collector's Society. Marietta, Ohio: The Glass Press Inc., 1995.

Measell, James, ed. *Imperial Glass Encyclopedia, Vol. II*. National Imperial Glass Collector's Society. Marietta, Ohio: The Glass Press Inc., 1997.

Measell, James, ed. *Imperial Glass Encyclopedia, Vol. III*. National Imperial Glass Collector's Society. Marietta, Ohio: The Glass Press Inc., 1999.

Measell, James and Wiggins, Berry. *Great American Glass of the Roaring 20s & Depression Era*. Marietta, Ohio: Antique Publications, 1998.

National Cambridge Collectors. *Colors in Cambridge Glass*. Paducah, Kentucky: Collector Books, 1984.

National Cambridge Collectors. *Reprint of 1930-34 Catalog*. Paducah, Kentucky: Collector Books, 1976.

National Cambridge Collectors. *Reprint of 1949 thru 1953 Catalog*. Paducah, Kentucky: Collector Books, 1978.

Nye, Mark, ed. *Reprint of 1940 Cambridge Glass Company Catalog*. Cambridge, Ohio: National Cambridge Collectors, 1995.

Pina, Leslie. *Depression Era Glass by Duncan*. Atglen, Pennsylvania: Schiffer Publications, 1999.

Pina, Leslie & Gallagher, Jerry. *Tiffin Glass 1914-1940*. Atglen, Pennsylvania: Schiffer Publications, 1996.

Schmidt, Tim. *Central Glass Works: The Depression Era*. Atglen, Pennsylvania: Schiffer Publications, 2004.

Scott, Virginia R. *The Collector's Guide to Imperial Candlewick*. Athens, Georgia, 1980.

Six, Dean. *Viking Glass 1944-1970*. Atglen, Pennsylvania: Schiffer Publishing, 2003.

Six, Dean. *West Virginia Glass Between the World Wars*. Atglen, Pennsylvania: Schiffer Publishing, 2002.

Six, Dean. Weston *West Virginia Glass, Monograph No.4*. Weston, West Virginia: West Virginia Museum of American Glass, 1999.

Smith, Bill and Phyllis, ed. *Cambridge Glass 1927-1929*. Springfield, Ohio: self published, 1986.

Torsiello, Paul & Debora and Tom & Arlene Stillman. *Paden City Glassware*. Atglen, Pennsylvania: Schiffer Publishing, 2002.

Townsend, Wayne. *Corn Flower: Creatively Canadian*. Toronto, Ontario: Natural Heritage Inc., 2001.

Vogel, Clarence W., *Heisey's Art and Colored Glass, 1922-1942*. Plymouth, Ohio: Heisey Publications, 1970.

Walker, William P., Bratkovich, Melissa, and Walker, Joan C. *Paden City Glass Company 1916-1951*. Marietta, Ohio: Antique Publications, 2003.

Weatherman, Hazel Marie. *Colored Glassware of the Depression Era 2*, Springfield, Missouri: Weatherman Glassbooks, 1974.

Weatherman, Hazel Marie. *Fostoria the First 50 Years*. Springfield, Missouri: Weatherman Glass Books, 1974.

West Virginia Museum of American Glass. *Archival Material*. Weston, West Virginia: West Virginia Museum of American Glass, 2001.

Wetzel-Tomalka, Mary. *Candlewick The Jewel of Imperial*. Marceline, Missouri: Walsworth Publishing Company, 1995.

Wilson, Charles. Westmoreland Glass. Paducah, Kentucky: Collector Books, 1996.

Index of Corn Flower Shapes